T0181756

Studies in Computational Intelligence

Volume 770

Series editor

Janusz Kacprzyk, Polish Academy of Sciences, Warsaw, Poland
e-mail: kacprzyk@ibspan.waw.pl

The series "Studies in Computational Intelligence" (SCI) publishes new developments and advances in the various areas of computational intelligence—quickly and with a high quality. The intent is to cover the theory, applications, and design methods of computational intelligence, as embedded in the fields of engineering, computer science, physics and life sciences, as well as the methodologies behind them. The series contains monographs, lecture notes and edited volumes in computational intelligence spanning the areas of neural networks, connectionist systems, genetic algorithms, evolutionary computation, artificial intelligence, cellular automata, self-organizing systems, soft computing, fuzzy systems, and hybrid intelligent systems. Of particular value to both the contributors and the readership are the short publication timeframe and the world-wide distribution, which enable both wide and rapid dissemination of research output.

More information about this series at http://www.springer.com/series/7092

Andrzej Bielecki

Models of Neurons
and Perceptrons: Selected
Problems and Challenges

 Springer

Andrzej Bielecki
Faculty of Electrical Engineering,
    Automation, Computer Science
    and Biomedical Engineering
AGH University of Science and Technology
Cracow
Poland

ISSN 1860-949X   ISSN 1860-9503   (electronic)
Studies in Computational Intelligence
ISBN 978-3-030-07942-0   ISBN 978-3-319-90140-4   (eBook)
https://doi.org/10.1007/978-3-319-90140-4

Printed on acid-free paper

This Springer imprint is published by the registered company Springer International Publishing AG
part of Springer Nature
The registered company address is: Gewerbestrasse 11, 6330 Cham, Switzerland

Contents

1 Introduction 1

Part I Preliminaries

2 Biological Foundations 7

3 Foundations of Artificial Neural Networks 15
   3.1   Models of Neurons and Synaptic Transmission............ 16
   3.2   Artificial Neural Networks and Their Applications 17
        3.2.1   Taxonomy of Neural Networks.................. 18
        3.2.2   Taxonomy of Training Methods 20
        3.2.3   Applications of Neural Networks 21

Part II Mathematical Foundations

4 General Foundations 31

5 Foundations of Dynamical Systems Theory 35
   5.1   Preliminaries.. 35
   5.2   The Euler Method on a Riemannian Manifold 39
   5.3   Linear Dynamical Systems 41
   5.4   Weakly Nonlinear Dynamical Systems................... 42
   5.5   Gradient Dynamical Systems.......................... 43
   5.6   Topological Conjugacy 43
   5.7   Pseudo-orbit Tracing Property 49
   5.8   Dynamical Systems with Control....................... 51
   5.9   Bibliographic Remarks 54

Part III   Mathematical Models of the Neuron

6   Models of the Whole Neuron   59
    6.1   Bibliographic Remarks   65

7   Models of Parts of the Neuron   67
    7.1   Model of Dendritic Conduction   67
    7.2   Model of Axonal Transport   69
    7.3   Models of Transport in the Presynaptic Bouton   71
        7.3.1   The A-G Model of Fast Transport Based on ODEs   71
        7.3.2   The Model of Fast Synaptic Transport Based
            on PDEs   73
        7.3.3   Model of Neuropeptide Slow Transport   83
    7.4   Model of the Synapse   92
    7.5   Bibliographic Remarks   95

Part IV   Mathematical Models of the Perceptron

8   General Model of the Perceptron   99
    8.1   Model of a Structure of a Neural Network   99
    8.2   Supervised Deterministic Training Process 106
    8.3   Gradient Learning Process 108
    8.4   Bibliographic Remarks 110

9   Linear Perceptrons 111
    9.1   Basic Properties of Linear Perceptrons 111
    9.2   Dynamics of Training Process of Linear Perceptrons 113
    9.3   Stability of the Learning Process of Linear Perceptrons 117
    9.4   Bibliographic Remarks 119

10   Weakly Nonlinear Perceptrons 121
    10.1   Bibliographic Remarks 124

11   Nonlinear Perceptrons 125
    11.1   Bibliographic Remarks 132

12   Concluding Remarks and Comments 133

Part V   Appendix

13   Approximation Properties of Perceptrons 137
    13.1   Bibliographic Remarks 139

14   Proofs 141
    14.1   Estimation of Constants in Fečkan Theorem 141
    14.2   Estimation of the Euler Method Error on a Manifold 144

References 149

Chapter 1
Introduction

In contemporary natural sciences strong mutual relations can be observed. Thus, biology is associated with chemistry and physics. It is connected, furthermore, with computer science, electronics, mathematics, cybernetics and philosophy which contribute significantly into biological studies. Let us specify the aforementioned relations in detail - see Fig. 1.1.

Biology, as such, treats of structures and processes in living systems. The two latter are studied by using observations and experiments, often sophisticated and technologically advanced. The obtained results are, on the one hand, the starting point for biological theories, for instance, to the paradigm that specific features are inherited genetically. On the other hand, in the contemporary biology, formal models of structures and processes are created. In order to create an adequate model, the properties which are crucial for the modelled phenomenon have to be specified. The semi-formal description, which is in its character, in a way, analogous to axiom systems in mathematics, is introduced on the basis of the specified properties - see, for example, [41], Sect. 2. This description is the starting point for creating either a formal model (the arrow 4 in Fig. 1.1), for instance mathematical one, or some implementations. The aforementioned implementations can have two forms - of a software algorithm or an electronic system. The latter one should be functionally similar to the modelled process or structure (the arrow 5). There is a reciprocal relation between a formal model and its implementation - each one can be the starting point for the other. For instance, if the ordinary differential equation, which models the dynamics of the studied biological process, can be easily obtained on the basis of the semi-formal description, then the electronic circuit, whose dynamics is described by this equation, can be constructed (the arrow 6). On the contrary, if the structure of the electronic circuit can be derived directly from the semi-formal description, then this circuit can be implemented and the differential equation, which describes its dynamics, can be derived (the arrow 7). Regardless of the order of the model and the implementation creation, the model can be analyzed by using formal approach (the

© Springer International Publishing AG, part of Springer Nature 2019
A. Bielecki, *Models of Neurons and Perceptrons: Selected Problems and Challenges*, Studies in Computational Intelligence 770,
https://doi.org/10.1007/978-3-319-90140-4_1

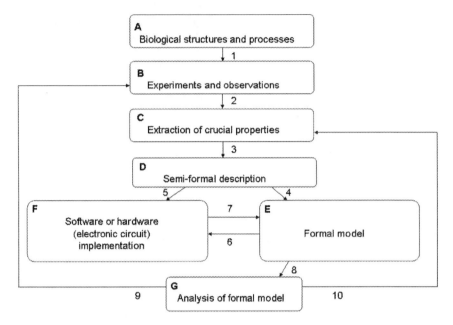

Fig. 1.1 The general schema of relations in modelling biological structures and processes

arrow 8). This analysis allows the researcher to study the properties of the formal model and, as the consequence, the properties of the investigated phenomenon. The results of this analysis can point out the directions for further observations and experiments (the arrow 9) as well as the necessity to modify the set of crucial properties (the arrow 10). In such a way the formal models and both software and electronic implementations become a significant part of the methodology of biological studies and generate specific methodological and philosophical problems [34].

This monograph treats of models of neural networks in the context of their modelling. The artificial systems, which are modelled after biological neural cells and structures constituted by them, are created for two reasons. On the one hand, such an approach enables the researchers to study biological phenomena indirectly by investigating artificial models. On the other hand, artificial neural networks (ANNs, for abbreviation) are computational systems of artificial intelligence. The mentioned systems enable researchers to solve a wide class of problems - pattern recognition, control, classification and diagnostics can be put as examples. Modelling of neural systems, hardware and software implementations of these models and, first of all, analysis of the models by using mathematical tools is the main topic of this monograph. Thus, referring to Fig. 1.1, problems which correspond to the frames E, F, G and partially, D, as well as the relations symbolized by the arrows 4, 5, 6, 7 and 8, are the topics of this monograph. It should be stressed, however, that only some selected problems are discussed.

In scientific investigations, the studies concerning mathematical modelling of the biological neural structures and artificial neural networks are, usually, regarded as separate topics. Nevertheless, all types of ANNs, as well as their training algorithms, are based on the models of biological prototypes. Therefore, in this monograph, the intension of the author is to unify these two topics. The more so because it seems that there are numerous models of biological neural structures that can be the basis for artificial systems and that have not been utilized yet. The way of the problem presentation in this monograph can be prospective for the abovementioned reasons.

This monograph consists of five parts. The first part, the preliminary one, treats of the foundations of both neuroscience (Chap. 2) and ANNs (Chap. 3). It should be stressed that biological foundations are presented more detailed than usually in the books that concern neurocybernetics and neuromathematics. In Chap. 3 all types of ANNs are discussed. In the second part (Chaps. 4 and 5) foundations of mathematical tools used in the sequel are specified. Chapter 4 treats about mathematical foundations. It deals with basic issues and as such can be omitted by mathematicians. In Chap. 5 very special topics of dynamical systems theory are presented and it can be interesting even for the professional mathematicians. Mathematical models of the neuron, both the whole one (Chap. 6) and its parts (Chap. 7), are discussed in the third part of the monograph. The models discussed in Chap. 7 are based on differential equations and they describe the processes of signal transmission inside of the neuron. Electronic implementations of these models are discussed widely as well. The Sects. 7.3.2 and 7.3.3 present the results obtained by the author. They concern fast and slow transport phenomena in the presynaptic bouton. The mathematical models of the perceptron are discussed in the fourth part. This part of the monograph also refers to the results obtained by the author. In Chap. 8 the model of the perceptron structure, as well as the general model of gradient training process of the perceptron, is presented. Dynamical aspects of training process of linear, weakly nonlinear and nonlinear perceptrons are analyzed in Chaps. 9, 10 and 11, respectively. The analysis is based on the dynamical systems theory and refers to the stability of a dynamical system, the flow discretization, the topological conjugacy of cascades and the shadowing property. Concluding remarks are presented in Chap. 12. Appendix (Chaps. 13 and 14) is the fifth part of the book. The dynamical models are the topic of the monograph. The approximation capabilities of perceptrons, however, are the very classical and well-worked topic in mathematical analysis of perceptrons. Thus, the basic and classical results are presented in Chap. 13. In the text of this monograph only the proofs of the theorems that concerns the topis directly are presented. The proofs of other theorems that are not known widely but has been utilized in this monograph, are presented in Chap. 14.

Part I
Preliminaries

Chapter 2
Biological Foundations

Each type of biological cells, including the simplest bacteria, receives stimuli from its environment and processes the obtained signals. Nevertheless, only metazoans, are the group whose representatives are equipped with neurons - the cells highly specialized in signal processing and transmission. In particular, only neurons are able to transmit signals over long distances. Neurons constitute multicellular structures including the most complex one which is the brain that is built not only from neurons but also from glia cells. The evolutionary development of the brain enable animals to perform intensional, not only reflexive, movements. Each consciously and intentionally initiated movement has its origin in the centers of movement control. *"For this reason, the motoneurons are arranged in the neocortext according to the body parts they innervate. Basal ganglia and the cerebellum are connected with the neocortex by extensive nerve tracts and build separate feedback loops for control and estimation of the outcome of the planned actions from the neocortex. (...) Voluntary movements and goal-oriented movements, which need sensory guidance, are not possible without sensory control. This exemplifies that, with the enlargement of the cortex, especially the more flexible, goal oriented and (...) more autonomous movements become possible"* [162], Sect. 8.3. The development of the brain enables living organisms to create complex models of the surrounding world and, as a consequence, to predict and plan events as well as achieve the goals that were planned beforehand. All of this allows the individuals equipped with complex central nervous system, first of all primates, to be a fully autonomous systems in the sense studied in [33, 134, 135, 161–163]. Let us recall briefly some basic facts concerning neural cells, the structures created by them and their functional properties.

Although there are a few types of neurons - see, for instance, [166], Chap. 5 - they have, in general, both the same structural scheme (see Fig. 2.1) and functional properties.

© Springer International Publishing AG, part of Springer Nature 2019
A. Bielecki, *Models of Neurons and Perceptrons: Selected Problems and Challenges*, Studies in Computational Intelligence 770,
https://doi.org/10.1007/978-3-319-90140-4_2

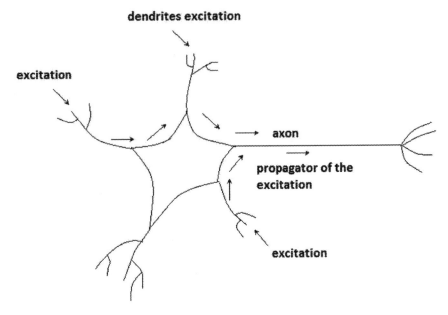

Fig. 2.1 The general schema of a neuron

Dendrites, the cell body and the axon are usually specified as the main parts of the neuron if it is regarded as the unit which transmits and processes signals. Similarly to all other types of biological cells, the neuron is separated from its environment by the cellular membrane. The membrane is a lipid bilayer which is an insulator but, because of the presence of electric channels and pumps which enables passive and active ion transport across the membrane, it is electrically polarized. In the neuron baseline state there is a difference between the electric potential of an internal and external sides of the membrane - the so called resting polarization. The internal side has negative value of a potential. When a neuron is excited, then the resting polarization is distorted at the point of excitation. This distortion is propagated along the neuron membrane as a wave. The stimulus can be of mechanical, electrical or chemical character. It should be stressed, however, that the statement that the neuron simply transmits signal from a dendrite to the synaptic bouton is an overmuch simplification. The neuron is stimulated mainly via dendrites but other parts of a cell can be stimulated as well. Furthermore, the signal is processed during propagating, for instance at the dendritic forks. The way it is processed in these spots depends on the geometry of the fork, among others on the relative lengths of diameters of its branches as well as it depends on the strength of the signal in both branches of the fork - see [166], Chap. 5. Furthermore, the neural cell body has low excitatory threshold and therefore it is sensitive to even low signals which can be propagated from synapses of other neurons which lie near the cell. The aforementioned wave is propagated along the membrane and stimulations from various parts of a neuron, processed during propagation, are summarized in the axon. Signal transmission to

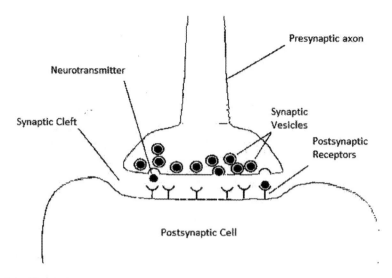

Fig. 2.2 Chemical synapse - a general scheme

the neuron terminal is the main function of this part of the neuron. The axon is myelinated. The myelin is an insulator but there are gaps between myelin segments in which additional stimulations from other neurons can be received and to act as control signals. The synaptic bouton, which is a neuron terminal, is the first part of the synapse, that makes the connection between two neurons. The synapse consists of three parts: the aforementioned presynaptic terminal bouton, synaptic cleft and the input of the postsynaptic dendrite of the neighbouring neuron - see Fig. 2.2.

The molecules of neurotransmitters, such as noradrenaline and acetylcholine, are synthesized and packed to vesicles in the terminal bouton. The vesicles have various sizes and, as a consequence, they contain various numbers of neurotransmitter molecules - from 3000 in small vesicles to over one million in the large ones [184]. The vesicles move inside of the terminal bouton probably utilizing cytoskeleton as paths along which transport takes place. Some of vesicles are docked in the specific region of the cellular membrane. Vesicles do not leave the domain unless the action potential arrives. The arrival of the action potential, i.e. the wave of the membrane polarization, open voltage gated channels and, then, the membrane ion pumps drive calcium ions inside of the bouton. This process triggers the release of neurotransmitter from docked vesicles through some period of time - see Fig. 2.3. The number of vesicles that release their content into the synaptic cleft is proportional to the vesicle concentration in the vicinity of the release site. It should be also stressed that exocytosis can take place in two ways according to the way in which the vesicle interacts with the cell membrane. In the full fusion the vesicles collapse into the plasma membrane and, as a consequence, the whole content of the vesicle is released to the synaptic cleft. In the kiss-and-run fusion only a part of the content of a vesicle is released

Fig. 2.3 Chemical synapse - transport of neurotransmitters in a presynaptic bouton. Neurotransmitter is packed to vesicles (1). Then, the vesicles diffuse towards the cell membrane (2) where they dock (3). When the action potential arrives, voltage gated channels open (3) and membrane ion pumps drive calcium ions inside the bouton. In the response to the inflow of the ions (4) the neurotransmitter is released from docked vesicles to the synaptic cleft (5)

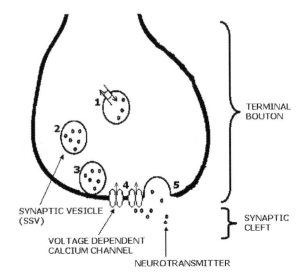

[184]. Such mechanism can control precisely the neurotransmitter signalling, among others it can retrie vesicles with great fidelity [159].

Depending on the part of the neuron to which the axon terminal is connected three types of synapses can be specified: axodendritic, axosomatic and axoaxonic [170], Chap. 5. In the first one the axon is connected to the dendrite of the postsynaptic neuron and this type of a synapse acts as a signal transmitter between two neurons. The two other types of synapses, in which the axon terminal is connected to the soma or to the axon of the other neuron, act probably as the modulators of a transmitted signal. It should be also mentioned that in neural networks dendrodendritic interactions also occur [170, 171].

It should be stressed that neurons communicate chemically by using two mechanisms, recognized as fast and slow synaptic transmissions. Fast transport of neurotransmitters, described briefly above, consists in stimulation a target cell within milliseconds whereas neuropeptide interaction lasts even several minutes. Neuropeptides are synthesized in the body of a neuron. Then, they are packed there in vesicles (so called large dense core vesicles - LDCV for abbreviation) and sent to the presynaptic bouton. Neuropeptides are activated by calcium ions. Space distribution of the inactive LDCVs is not uniform and their diffusion is slow, whereas the activated LCDVs diffuse faster and their diffusion is undirect. Only the activated LDCVs can be released to the synaptic cleft - see Fig. 2.4. Slow-acting neurotransmitters control the efficacy of the fast synaptic transmission by regulating the efficiency of neurotransmitter release from presynaptic terminals and by regulating the efficiency with which the fast-acting neurotransmitters produce their effects on postsynaptic receptors [85]. More detailed discussion concerning the slow transport of neuropeptides can be found in [125].

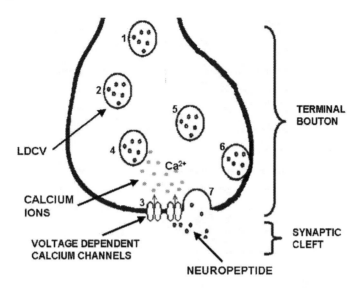

Fig. 2.4 Chemical synapse - transport of neuropeptides in a presynaptic bouton. The large dense core vesicles are filled with neuropeptide before they arrive from the cell soma (1). Then, they are captured and immobilized (2). The voltage gated channels are opened by the action potential (3) and ionic pumps drive calcium to the cytoplasm of the bouton. There, the calcium activates immobile vesicles (4). After diffusing to the cytoplasm (5) the active vesicles arrive in the vicinity of the membrane (6). Then, they are docked to the cell membrane and the neuropeptide is released (7)

In both types of the synapses described above the interaction has chemical charac-ter which means that chemical substances - neurotransmitters and neuropeptides - are sent between two neurons. In such synapses, called chemical synapses, the synaptic cleft has the length of $20 \div 40$ nanometers. In the electrical synapses (gap junc-tions) the length of the synaptic cleft is equal to $1.2 \div 3.5$ nm. Such small distance between two neurons enables the ions to flow by using the gap junction channels. In the electrical synapses the signal control is not as precise as in the chemical synapses but signal transmission is significantly faster [22]. Therefore the electrical synapses are used to trigger fast reactions. Furthermore, the electrical synapse acts as a low pass filter [22]. Moreover, the electrical synapse can be regarded as a synchronizing element - for instance a few networks constituted by electrical coupling were found in the neocortex [92, 168, 169].

In the central nervous system the signal is propagated not only from a presynaptic neuron to a postsynaptic one but also the backward signalling, based on depolariza-tion induced suppression of inhibition (DSI) phenomenon is observed. This mecha-nism utilizes the transport of endocannabinoids from the postsynaptic neuron to the presynaptic cell. The gamma-aminobutyric acid (GABA) is released and that causes stopping a neurotransmitter transmitting. The DSI mechanism enhances the long-term potential ([142, 185]) - it was shown experimentally that endocannabinoids extinguish the negative emotions triggered by the reminders of past experiences, by using control activity of GABA - see [120] and [121], Chap. 8.

It should also be mentioned that the signal processing in neural networks has the analog-digital character and both aspects have to be considered. Both the subliminal analog potentials that are transmitted non-synaptically in the nervous network - see discussion below concerning signalling in extra-cellular space (ECS, for abbreviation) - and digital spiking discharges of the whole neuron have to be taken into account, which was postulated in the 70' of the 20th century as one of the neurophysiological foundations of psychiatry [107], p. 185.

As it has been mentioned above, the signals in neural networks can be propagated non-synaptically by diffusion in ECS. The active substances can diffuse to neurons, glial cells and capillaries without using synapses. This type of the signal processing can function between both neurons and glia. It seems that it is the basis for integrating of the signal processing between distant cells and it involves large numbers of units [9, 174].

The aforementioned glial cells play key roles in the nervous system. They constitute the scaffolding on which the neural network is stretched and they insulate one neuron from another. They also take part in metabolism by supplying oxygen and nutrients to the neurons. Furthermore, they protect neurons from pathogens. They also take active role in repairing the damaged neural network. According to the recent studies it turns out that glia play an important role in neurotransmission. For instance, the release of ATP, among others from glia, activates the membrane receptors that modulate intracellular calcium. By using this mechanism glia not only detect neural activity but also communicate among other glial cells [70]. Not only glia-glia communication was observed but also signal transfer between glia and neurons is possible. The signal transmission between neurons and glial cells is realized by using chemical conduction, ion fluxes and cell adhesion molecules [72]. The communication between neurons and glia has significant influence on homeostasis of the neural processes by regulation the synaptic strength, gene expression, mitotic rate and differentiation of cells in the dependence of the activity in neural network [71]. Glia activity is also crucial for learning process as well as for forming long-term memory because they take part in forming synapses [9, 69, 174].

Neurons and glia constitute network of interacting cells. In the neural network some areas can be distinguished according both to they roles in signal processing and to the structure which is generated by connections between neurons. In general, the following structures of neurons can be listed: multilayer structures, recurrent structures and local connections. In multilayer structures neurons form layers in such a way that the neurons of a given layer are connected only to the neurons of the next layer. Such connections are characterized for these part of neural network which process signals from senses, first of all from sight and hearing. In the recurrent structures the stimulation of some neurons causes excitation in closed loops. In such a way, after a single external stimulation a neuron in recurrent network is usually stimulated several times. In the neural structures organized as local connections the neurons are located in the nodes of two-dimensional or three-dimensional lattice. The structure of the lattice determines how many neighbours each neuron has. The neuron communicates with its neighbouring neurons. In such networks the external stimulation is propagated through the whole network as a wave of excitations.

The structures and phenomena briefly discussed above are starting points for mathematical, computer science, cybernetic and electronic models. Very often a given model has more than one of the listed aspects. The processes in subregions of neurons, for instance excitation of neural membrane, axonal transport, transport in presynaptic bouton, synaptic conductance as well as synaptic plasticity, are modelled by using both extremely simple and sophisticated tools. Some of these models are starting points for the studies concerning artificial neural networks. These problems are considered in Chap. 6 in reference to single neuron and in Chaps. 8, 9, 10 and 11 in reference to perceptrons. It should also be mentioned that although glial cells have not been taken into consideration in artificial neural networks, the possibilities of taking into account them in neural network model have been already discussed [49].

Chapter 3
Foundations of Artificial Neural Networks

The rapid growth of computational power of computers is one of the basic qualities in the development of computer science. Therefore, informatics is applied to solving more and more complex problems and, what follows, the demand for bigger and more complex software occurs. It is not always possible, however, to use classical algorithmic methods to create such a type of software. There are two reasons for it. First of all, a good model of the relation between the input and output parameters often either does not exist at all or it cannot be created at the present level of scientific knowledge. It is worth of mentioning that the algorithmic approach requires the knowledge of the explicit form of the mapping between the aforementioned sets of parameters. Secondly, even if the model is given, the algorithmic approach can be impossible regarding its over-complexity. It can be both complexity of the task on the stage of the algorithm creating, and too slow working of the implemented system. The latter one is a critical parameter especially in the on-line systems. Therefore, the alternative approaches, in comparison with the classical algorithmic approach, are developed intensively. Artificial neural networks are included into this group of methods.

The neurophysiological studies of functional properties of nervous systems enabled researchers at the beginning of the 1940's to formulate the cybernetic model of the neuron [58] which, slightly modified, is commonly used up to present. At the turn of 1950's and 1960's the first artificial neural systems - PERCEPTRON and ADALINE were constructed. They were electromechanical systems. The first algorithms for setting of the synaptic weights in such a type of systems were worked out. Those pioneering attempts attracted attention to the possibilities of such systems. At the same time, however, significant limits were discovered. Nowadays, from the perspective of the time, it is known that on the one hand, the limits were caused by the lack of proper mathematical models of neural networks. On the other hand, they were caused by the application of just one type of artificial neural networks - the multilayer

© Springer International Publishing AG, part of Springer Nature 2019
A. Bielecki, *Models of Neurons and Perceptrons: Selected Problems
and Challenges*, Studies in Computational Intelligence 770,
https://doi.org/10.1007/978-3-319-90140-4_3

ones. They consisted of binary neurons and were the only ones known then. Nevertheless, the heated criticism of the new approach, first of all [138], caused more or less 15-year-long lasting impasse in the research on artificial neural networks. There came a breakthrough of the half of 1980's when many new types of artificial neural networks and their training algorithms were introduced. Mathematical analysis of artificial neural systems was initiated then as well. That is one of two main topics of this monograph. The formal models of biological neurons is the second one. In this monograph the models based on differential equations and dynamical systems are considered. Electronic circuits that reflect some functions of neurons are considered as well.

3.1 Models of Neurons and Synaptic Transmission

The aforementioned cybernetic model, founded by McCulloch and Pitts, is the simplest possible model. According to this approach, the neuron is treated as an indivisible unit which realizes the stimulus-reaction scheme. Such types of models are generalizations of the model proposed by McCulloch and Pitts. In the deterministic models of such a type, the neuron is represented by a mapping dependent on the family of parameters whereas in the probabilistic models the neuron is modelled by a distribution of probability. In this monograph, various models of the neuron are discussed in Chap. 6.

Nevertheless various complex processes take place in the neuron. Even if the neuron is considered only as the unit which processes the signals received from its environment i.e., first of all, from other neurons and glia cells, a complex signal processing takes place in dendrites, in the axon and in the synapse - see Chap. 2. Thus, the neuron is considered as a complex unit and, what follows, the models that describe the processes that take place in various parts of the neuron, can be modelled individually by using mathematical tools. Thus, signal processing in dendrites, soma, axon and synapse is modelled separately. Since the chemical synapse consists of three parts - the presynaptic bouton, the synaptic cleft and the receptors of the postsynaptic dendrites - the processes in each part of the synapse can be modelled individually as well. The modelling of the whole chemical synapse that is considered as an indivisible module is an alternative possibility. The electric synapse, however, is structurally and functionally far simpler than the chemical one and therefore it is usually modelled as an indivisible module. The modelling of the cellular membrane conductance, including controlled conductance in various types of ionic channels, is another problem in the modelling of processes in various parts of the neuron.

Two types of mathematical models are used in the context of modelling signal processing in various parts of the neuron - the probabilistic models and the ones which are based on differential equations and dynamical systems theory. In the first ones the probabilistic aspect of the modelled processes is regarded as the most crucial. In the second ones, the dynamical aspect is regarded as the crucial one and, as a consequence, differential equation or a system of differential equations is used as

Fig. 3.1 Example of a perceptron which has one hidden layer

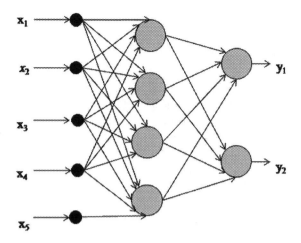

a model. Both partial differential equations and ordinary differential equations can be used as the mathematical tool. In the case of nonlinear models and most of the partial differential equations models it is impossible to find analytical solution of the problem. Therefore, numerical simulations are necessary which, in turn, generates subsequent both theoretical and practical problems concerning their stability, accuracy, convergence and computational complexity. If the model is based on ordinary differential equations then it can be implemented by using electronic circuit whose dynamics is described by the same differential equation. It should be also mentioned that, sometimes, it is more convenient to propose an electronic circuit functionally equivalent to a modelled process only on the basis of semi-formal description - arrow 5 in Fig. 1.1. Then, a formal model based on ordinary differential equation can be obtained as the model which describes directly the circuit - arrow 7 in Fig. 1.1. In this monograph the models of signal processing in neurons based on differential equations and dynamical systems theory are discussed. The possibilities of modelling neural phenomena by using electronic circuits are analyzed as well. Both analogue and digital electronic modules are taken into consideration and their relation with mathematical model are discussed in detail. This problems are described in Chap. 7.

3.2 Artificial Neural Networks and Their Applications

In this section the basic types of artificial neural networks are discussed in Sect. 3.2.1. Then, a review of training methods of neural networks is presented in Sect. 3.2.2 and applications of various types of artificial neural networks are presented briefly in Sect. 3.2.3.

3.2.1 Taxonomy of Neural Networks

Artificial neural network (ANN for abbreviation) is a system of artificial neurons that are connected in such a way that output signals of neurons are put onto inputs of the other ones. Taxonomy of ANNs can be done according to various criteria. The type of neurons that constitute an ANN can be one of such criterion. Thus, an ANN can consist of McCulloch-Pitts type neurons, radial neurons or neurons with memory including neurons with hysteresis. The aforementioned neurons are equivalent to the mappings according to which the input stimulus is transformed in the deterministic way and it generates the output signal. Probabilistic neurons are another class of artificial neurons. In such neurons if an input signal is put then the output signal is generated by using a given probabilistic distribution. Various types of artificial neurons are described in details in Chap. 6.

The structure which is constituted by connected neurons can be another criterion of ANNs taxonomy. In such a context three following types of neural networks can be specified.

Multilayer neural networks.   In this type of ANNs, called also perceptrons, the signal is propagated in one direction. Neurons constitute layers in such a way that each neuron of a given layer is connected only with all neurons of the next layer - see Fig. 3.1. A perceptron has the input units and the output layer. It usually has one hidden layer or, in some cases, more. It is caused by the fact that such the structure with one hidden layer is sufficient to approximate any continuous mapping on a compact set. This problem is discussed briefly in Chap. 13. If an input vector **x** is put onto input units, then each neuron of a given layer is stimulated only one time, and it propagates the signal onto inputs of the neurons of the next layer. As the result, the output signals are generated on outputs of the output layer. These signals constitute an output signal which is a vector **y**. In such a way a perceptron creates an output which is the answer to the stimulation **x**.

Recurrent neural networks.   Neural connections in recurrent ANNs form loops. Therefore, after stimulation of a group of neurons by an input signal **x**, the neurons are stimulated repeatedly and they generate dynamical process of the neural network excitation. After some time a recurrent network, activated by a signal **x**, can achieve a stable state which means that in two successive iterations the output signals of all neurons do not change. As a consequence, the states of all neurons remain unchanged after all following iterations. Oscillations of the states of neurons or chaotic dynamics are another possibilities. Hopfield networks, in which each neuron is connected with all neurons in the network, are the basic type of recurrent ANNs - see Fig. 3.2.

Cellular neural networks.   Neural connections in cellular neural networks are local. Neurons are placed in the nodes of the net whose structure is defined. It usually is a square net on a plane but other structures, including three-dimensional, are also possible - see Fig. 3.3. A neuron is connected only with neighbouring ones and a neighbourhood is defined according to the structure of the net. Since the neurons situated on the edge of the net disturb the symmetry, the edges of the

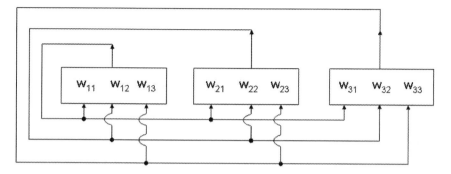

Fig. 3.2   The schema of a Hopfield neural network which consists of three neurons

Fig. 3.3   The schema of a
cellular neural networks with
square neighbourhood (top)
and hexagonal
neighbourhood (bottom). If
in the square network the
additional connections
between the following pairs
of the neurons: A-D, E-H,
I-L, M-P and A-M, B-N,
C-O, D-P were created then
the toroidal structure of the
network connections would
be obtained. In the analogous
way, adding connections
between pairs: A-D, E-I, J-M
and A-J, B-K,C-L, D-M in
the hexagonal network leads
to the toroidal structure of
the network geometry

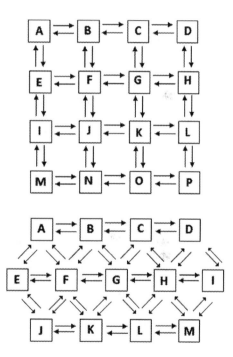

net that define the structure of a cellular neural network can be identified and, in
such a way, a toroidal net is obtained. In cellular networks a stimulation is propagated in the form of a wave. Its dynamics is described by a partial differential or
difference equation.

The types of artificial neural networks specified above are the basic ones. There
are other several types of ANNs. Apart form the aforementioned ART-type networks, Hamming neural networks, Kohonen networks or RBF networks can be put
as examples.

3.2.2   Taxonomy of Training Methods

If an artificial neural network is planned to be used for solving a given problem, then parameters of its neurons have to be set. The process of setting the parameters is called a training or a learning process. There are numerous methods of setting of the parameters. Let us present the taxonomy of training processes.

1. **Supervised learning**

   In a supervised training pieces of information about a properly trained neural network are given. This information is used in the training process. The following supervised training strategies can be specified in dependence on this a priori given information character and the way the information is used during the training process.

   (a) **Iterative methods** are common types of supervised training algorithms. In these methods parameters of neurons are changed iteratively and changes of parameter values in a single step are small. In the group of iterative training algorithms two subgroups can be specified.

      i. In **delta-methods** a set of input signals, for which the correct output signals of the network is known, is given. Thus, it can be specified a so-called *training set* which is a finite sequence of pairs $\{(\mathbf{x}^{(n)}, \mathbf{z}^{(n)})\}_{n\in\{1,\dots,N\}}$, where $\mathbf{z}^{(n)}$ is a signal which should be an output one if the input signal $\mathbf{x}^{(n)}$ is put onto the input of the neural network. The untrained ANN, however, generates an output signal $\mathbf{y}^{(n)}$, usually different from $\mathbf{z}^{(n)}$. Then, the difference between these two output signals, the current one $\mathbf{y}^{(n)}$ and the proper one $\mathbf{z}^{(n)}$, is measured by using so-called *criterial function* and the neuron parameters are modified slightly in order to obtain smaller difference. The modification of the parameters is done in a deterministic way, commonly by using a differential method, often a gradient one. In such a case the criterial function has to be differentiable. It should be mentioned that the analysis of a training process dynamics of perceptrons generated by a gradient method is one of the main topic of this monograph.

      ii. **Probabilistic methods** consists in probabilistic setting of the parameters of neurons according to the given probabilistic distributions. The distributions are obtained on the basis of the pieces of information that concern the proper solution. This setting of parameters is an iterative process as well. Genetic algorithms, sometimes used as the training strategies of neural networks, can be put as examples of these types of methods.

   (b) **Non-iterative methods** have nowadays marginal significance. They can be used if the problem is simple and neural network is small. In such a case it is sometimes possible to find the proper values of parameters by using theoretical methods. In the simple case, for instance, weights of McCulloch-Pitts-type neuron can be calculated as the solutions of a system of linear algebraic inequalities. In Hopfield neural networks, in some sorts of problems, an algebraic formula for calculation of weights values can be inferred from theoretical considerations as well.

2. **Unsupervised learning**

   In this group of training methods no information about the solution is given a priori. Thus, the training set contains only some input signals without any reference to the proper answers of the ANN. According to the basic approach to the training process, unsupervised training methods can be divided into two subgroups.

   (a) **Competitive methods** consist in selection of the neuron or a group of neurons that have been strongest stimulated by a given input signal. Then, only the weights of this neuron (the *winner takes all* algorithm) or of this group of neurons (the *winner takes most* algorithm) are modified in a current step of the training process. This type of training methods are used, first of all, in Kohonen networks. This group of methods as well as the Kohonen networks can be used for detection of relations in the set of input signals, first of all for clusterization of the input signal set. As the result of the training process the neurons represent clusters of input signals.

   (b) **The methods based on the Hebb's rule,** according to which a neuron stimulated repeatedly in short time becomes more and more sensitive to this type of stimulation: *"Let us assume that the persistence or repetition of a reverberatory activity tends to induce lasting cellular changes that add to its stability. [...] When an axon of cell A is near enough to excite a cell B and repeatedly or persistently takes part in firing it, some growth process or metabolic change takes place in one or both cells such that A's efficiency, as one of the cells firing B, is increased."* [89]. The creation of the patterns of stimulations of the neurons in a network in the response to the input signals is the result of the training process.

3.2.3 Applications of Neural Networks

As it has been aforementioned, the mathematical foundations of perceptrons is one of the crucial topics of this monograph. Artificial neural networks, however, offer such wide possibilities of applications that they should be mentioned in the introductory part of this book. Let us discuss the specificity of ANNs in the context of their applications. Then, examples of the applications will be presented.

Artificial neural networks belong to the class of distributed connectionist systems (see, for instance, [78], Chap. 3) in which the knowledge is distributed among the units that constitute the AI system. In ANNs the knowledge is encoded as the parameters of neurons - the weights in the case of McCulloch-Pitts neurons. The parameters are set automatically during a training process. In the trained neural network the knowledge is encoded implicitly which means that apart from very specific types of

ANNs the parameters of the neurons have not any direct interpretation which could be the basis to decode the knowledge. Therefore, at the cognitive level, the causal relation between the stimulus and the network reaction cannot be traced. This is one of the crucial drawbacks of artificial neural networks. On the other hand, however, in order to apply a neural network to solve a problem, it is necessary to know only the factors the solution is dependent on. Therefore, artificial neural networks can be applied successfully to the problems for which the effective algorithm cannot be worked out directly because of the high level of complication or because of the fact that the cause and effect chain between the inputs and the solution remains unknown. Since the input of neural networks is of the vector form, the factors mentioned above have to be encoded as a vector or other mathematical structure that can be naturally interpreted as a vector, for instance a matrix. If the factors are numerical ones then they can be encoded in a natural way as the components of the input vector. If they have symbolic character, then the proper way of encoding them as components of a vector have to be found. It is crucial to reflect the relations between the symbolic data and not to produce relations that do not exist in the data set. For instance, if a day of the week should be taken into consideration as a part of the data encoded in the input vector, then the way of encoding should reflect cyclic character of this data. Representation the days of the week as seven points evenly distributed on a circle is an example of a proper solution. If, for instance, the letters of the alphabet are the input data, then do not exist any relations between them. Because it is impossible to create vector representation without any relations between the vectors, the relations between each two pairs of vectors should be the same. Encoding the letters as binary vectors in which only one bit is equal to one is an example of a proper solution. In a such way of the encoding, all vectors have the same structure - one active bit, and Hamming distance between each two different vectors is equal to two. Examples of such encoding can be found in [155], Sect. 4.1.1. The studies concerning sensitivity of neural networks to input data representation are presented in [154].

The aforementioned sensitivity is, among others, connected to the problem of input data transformation. If, for instance, one input parameter denotes the air temperature in Celsius degrees and the other one the atmospheric pressure in millimeters of mercury, then the first parameter can take values around zero whereas the second one oscillates around 760. The trained neural network, however, does not "interpret" the meaning of the parameters but only "experiences" them as a weak or strong excitement. Therefore, the temperature, as close to zero value, will not have any influence on the training process. Both parameters, however, should be treated as equally important. Therefore, in such cases, the transformation of parameters that constitute the input of the network is necessary. The input signal normalization is the most common used transformation of this type.

A normalization procedure corresponds to creating a mapping

$$F : \mathbb{R}^n \supset A \ni \mathbf{x} \rightarrow \hat{\mathbf{x}} \in \mathbb{R}^k, \text{ where } \|\hat{\mathbf{x}}\| = 1,$$

where A is the set of the input signals. The most commonly used normalization is done according to the formula $\hat{x} = \frac{\mathbf{x}}{\|\mathbf{x}\|}$. This formula defines projection,

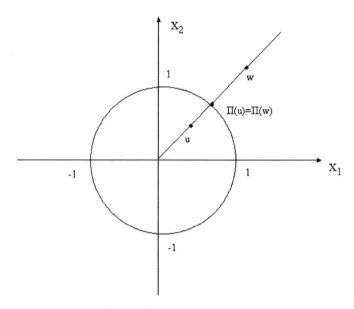

Fig. 3.4 Simple projection of $\mathbb{R}^2$ onto $\mathcal{S}^1$

$$\Pi : \mathbb{R}^n \setminus \{\mathbf{0}\} \to \mathcal{S}^{n-1} \subset \mathbb{R}^n,$$

let us call it a simple projection, of $\mathbb{R}^n \setminus \{\mathbf{0}\}$ onto $(n-1)$-dimensional sphere $\mathcal{S}^{n-1}$ - see Fig. 3.4 for the two-dimensional case.

A simple projection has crucial drawbacks. First of all, the dimension of the space is reduced. Secondly, the projection is not defined on the whole space because the mapping is undefined for the origin of the coordinate system. Furthermore, the space $\mathbb{R}^n$, which has an an infinite measure, is projected onto a sphere which has a finite measure. Additionally, the projection is not an injective mapping - if two points, let us say u and w, lie on the same radial line, then $\Pi(u) = \Pi(w)$ - see Fig. 3.4. Referring to the considered problem that means that if two data clusters are situated along the same radial direction then, after normalization, they cannot be separated even if they were well separated before normalization. Therefore, this method of input data normalization should be used only in such cases when is a priori known that clusters, in the space of input signals, are located in various radial directions.

The mentioned problems cause looking for the normalization which does not reduce the input signals space dimension. The stereographic projection

$$S : \mathbb{R}^n \to \mathcal{S}^n \subset \mathbb{R}^{n+1}$$

is an example of such a mapping. It was proposed as a normalization procedure for the data processed by neural networks - see [36]. Geometric interpretation of the stereographic projection is visualized in Fig. 3.5 for the two-dimensional case.

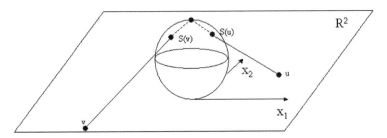

Fig. 3.5 Stereographic projection of $\mathbb{R}^2$

The stereographic projection transforms the $n-$dimensional Euclidean space into $n-$dimensional sphere that has its south pole in the origin of the coordinate system. The classical stereographic projection is given explicitly by algebraic formulae for each natural n. The formulae given below describes a little modified stereographic projection - the space $\mathbb{R}^n$ is transformed into $n-$dimensional unit sphere which has the centre at the origin of the coordinate system, as it is demand for the data processed by neural networks. Let $P = (x_1, ..., x_n)$ be a point in $\mathbb{R}^n$. Then $S(P) = \tilde{P} = (\tilde{x}_1, ..., \tilde{x}_{n+1})$ is given as

$$\tilde{x}_i = \frac{4x_i}{4+s} \text{ for } i = 1, ..., n;$$
$$\tilde{x}_{n+1} = \frac{s-4}{4+s},$$
(3.1)

where $s := \sum_{i=1}^{n} x_i^2$.

As it has been already mentioned, stereographic projection preserves the transformed space dimension and is defined on the whole $\mathbb{R}^n$. Furthermore, it is an injective mapping i.e. if $u \neq v$, $u, v \in \mathbb{R}^n$ then $S(u) \neq S(v)$. Nevertheless, it transforms a space which has an infinite measure into a space which has a finite measure. This implies, among others, that points that are far from each other in $\mathbb{R}^n$ can be closed each to other on $\mathcal{S}^n$. Therefore, two clusters which are well separated in $\mathbb{R}^n$ can be hardly separated after normalization. Such case, however, can only take place if the clusters are far from the coordinate system origin - then they are transformed near to the north pole of the sphere. Since, in practice, the norms of transformed vectors are limited, the minimal distance between clusters after signal normalization can be estimated. Having such estimation, the radius r of the sphere can be taken as large as possible because in normalization in the context of ANNs training process all input vectors should have the same norm, nor necessarily equal to one. In a such case formulae (3.1) have the following form

$$\tilde{x}_i = \frac{4r^2 x_i}{4r^2+s} \text{ for } i = 1, ..., n;$$
$$\tilde{x}_{n+1} = \frac{s-4r^2}{4r^2+s} \cdot r.$$
(3.2)

Artificial neural networks can be applied in many areas. The following applications can be put as examples. According to the topic of this monograph, the stress is put onto perceptrons and perceptron-like networks.

Industrial applications.   Possibilities of applications of various types of artificial neural networks in power industry have been widely studied. Two-layer perceptron was applied to the failure diagnostics of oil underground switches [65]. The status of the switch is characterized by the degree of corrosion of its internal tank. The parameters that reflected the status of the equipment such as the content of moisture in the oil, the parameter indicating loses of the oil and the dielectric strength were encoded as the components of the input signals. The output signal was binary: non-corrosive condition or corrosive condition. It turned out that the used perceptron classified the oil switch status significantly better than the commonly used the current threshold classification method. All the cases were classified correctly by the perceptron whereas the threshold method achieved 84% of efficiency. It should be mentioned, however, the the number of the classified cases was low and was equal to 19.

An interesting example of the application of an untypical neural network in power industry is described in [2]. A neural network which realizes a Parzen estimator [149] was used to power plant monitoring and diagnostics. This type of neural network was introduced by Specht [172]. Real sensor data obtained from the feedwater systems of an electric power plant were used as input for the Specht neural network. The neural network was integrated with influence diagrams in order to combine efficiency, economy and flexibility of the diagrams with learning abilities, parallel computation and the noise resistance of neural networks. The described method allows the operator to observe both the most likely failure causes and the probability ranking of them.
The intelligent monitoring of wind turbines by using ART-type neural networks is described in the series of publication [12–15, 35]. Various types of resonant networks as well as a hybrid system consisted of an ART network and RBF network were tested. The parameter describing operational states of a turbine and signals from vibration channels were put onto input of the systems. It turned out that the tested systems were managed to cluster operational states effectively in real times detecting the new ones that corresponded to faults.

There are numerous reasons for which the prediction of power load is one of the crucial problems in power industry. First of all, the electric production on industrial scale has great inertia, especially if it is produced by using the coal power stations. Secondly, in the selected types of power stations, first of all in the wind plants, production possibilities are hardly predictable. The prediction of twenty four hours load profile is a common task. Various types of artificial intelligent systems are used to solve this problem and ANNs are the ones that are applied universally. The single perceptrons, as well as hybrid systems that consisted of perceptrons and an expert module, were utilized successfully to predict the power load at the country level. In such systems the daily electricity usage, a day of the week, a day of the year and the weather temperature are used as the input parameters. The mean absolute percentage error (MAPE) of the prediction is contained at the interval $1.2 \div 5\%$ [19, 147]. It should be also mentioned that the other types of ANNs, not only perceptrons, were efficiently applied to the load term forecasting. Kohonen networks that are examples of a self-organizing systems achieved MAPE around 1% which is a very good result [108].

Engineering applications. There are numerous possibilities for application of various types of ANNs in engineering. Neural networks are widely used as control modules and cognitive systems that process, arrange and recognize sensor data including camera images. Furthermore, they are used as diagnostic systems.

The mentioned applications are commonly implemented in robotics. So called convolution neural networks (ConvNN) are very similar to perceptrons so they refer strongly to the topic of this monograph. This type of neural networks is dedicated to process and recognize camera images, often in the context of a robot vision system. It turned out that simple perceptrons are not a proper tool to solve this task because they need huge number of weights. Therefore, the convolution neural networks, in which layers are organized in a such way that this organization is optimal in the context of camera image processing, are the proper tools. Each layer of a ConvNN has its inner structure. The width and height of the input layer define the array of neurons equal to the number of pixels in the processed image whereas the depth of the layer corresponds to the number of colours. The subsequent layers, which also have the analogous structure, i.e. neurons are organized as three-dimensional arrays, process the image. The neurons of a given layer are connected only to a small region of the preceding layer. Thus, the inter-layer connections have a local character. The output layer reduces the input image into a vector encoding class scores. It should be also mentioned that the output layer is fully connected with the previous one, as in classical perceptrons. The effective application of a convolution neural network in a robot vision system is described in [124].

Medical applications. Artificial intelligence systems, not necessarily artificial neural networks, are widely used in medicine. There are two main areas of their application: medical imaging and medical diagnostics. In the context of medical imaging AI systems are used both for images processing and recognition and understanding [45, 143, 177]. Medical image understanding is strictly connected with medical diagnostics aided by computer systems. In such context syntactic method are used [25–27] as well as soft computing [177]. In the artificial intelligence systems applied in medicine the neural networks play an important role. Let us present a few examples.

The aforementioned convolution neural networks were applied in forensic dentistry [137]. In some cases dental records are crucial in forensic postmortem identification. Since the procedures of comparison of postmortem dental findings with antemortem records can be time consuming, especially in the case of huge disasters, the demand for automatization of the process occurs. In the cited paper the application of a deep convolutional neural network for the tooth types classification is described. A ConvNN is used for the process dental cone-beam computed tomography images and in order to recognize the type of a tooth. The system efficiency is equal to 91%.

In the paper [186] the application of a perceptron for diagnosis of skin diseases is described. The perceptron had one hidden layer consisted of 20 neurons and an output layer that had 10 neurons. The symptoms of diseases were encoded by using an input vector which had 96 components. The value of the component was equal to one if the symptom occurred, to zero if the symptom did not occurred and to $\frac{1}{2}$ in the case of the lack of information about the symptom appearance. The output signals corresponded to ten dermatoses. The accuracy of the diagnosis was equal to 70%.

The perceptron applications for recognizing of appearance of myocardial infarction is discussed in [16, 17]. The patients were presented to the emergency department with anterior chest pains. The symptoms were encoded in the analogous way as in the aforementioned perceptron for the diagnosis of skin diseases. It was only one neuron in the output layer. Initially, according to the medical knowledge, forty one symptoms were put onto the network input. The perceptron was trained by using the descent gradient method with momentum. It turned out that only twenty from among specified symptoms were necessary to make the perceptron act properly. There perceptrons with one and with two hidden layers were tested. The ability of the perceptron to distinguish patients with an acute myocardial infarction from those without the disease was compared with the diagnosis made by physicians. The best perceptron had two hidden layers, each consisted of ten neurons. It diagnosed correctly 92% patients with infarction and 95.7% patients without infarction. The diagnostic correctness for physicians was equal to 88% and 76% respectively.

Economic applications.   In economics artificial neural networks are used for prediction of time series in various contexts. Prediction of currencies rates, shares and bond values, prices on the markets as well as the demand and supply are the classical examples [10, 20, 158, 187]. In the paper [10] the USD to GBP exchange rate prediction by using a perceptron is described. Not only the previous values of the rate were used as the input signals but also the coefficient of random walk was taken into consideration.

A financial predictor implemented as a perceptron is described in [119]. The system predicts the value of the Dow Jones Industrial Average index for the next month. The index is one of the principal financial indices and it shows how thirty largest US companies have traded. Its values are calculated monthly. Twenty variables were put as the system input parameters: The changes of the index in three preceding months - three variables, the index of consumption prices in three preceding months - three variables, the prices of oil in three preceding months - three variables, the inflation rate in three preceding months - three variables, the interest rate in three preceding months - three variables, the unemployment rate in three preceding months - three variables, one variable which characterized the political situation and the last variable - the number of the month. In the 21st century the neural systems for the prediction of the Dow Jones Industrial Average index were studied intensively. The influence of various input parameter of the prediction accuracy is discussed in [131]. The authors also reported that the Root Mean Squared Error is equal to around 0.015.

Other applications.   Industrial, engineering, medical and economic applications of artificial neural networks are contemporary classical. Neural networks, however, can be applied efficiently also in other fields. Let us present one of them.

In each language a problem of phonematic translation appears. Phonematic transformation is a basic tool for any artificial system of speech synthesis. The task can be described as a translation of a written text into a string of phonematic characters that define the way in which a given letter should be pronounced. Phonematic transformation is a difficult task because of the context sensitivity. The transformation of a single letter into a phoneme is usually ambiguous because, in most cases, it depends on the characters both before and after the letter. In the context of phonematic translation some languages, for instance Italian, are very regular whereas others, for instance English, are extremely irregular. A neural system of the orthographic-phonematic translation for Polish language, which is rather regular in the context of this transformation, is described in a series of papers [18, 48, 155, 156]. The aforementioned problem of representation of symbolic data, which in this context are letters, appears in the phonematic translation task. Two representation were tested. In the unary representation a letter is encoded as a binary vector that has thirty two components according to the fact that there are thirty two letters in the Polish alphabet. Only one component in the vector, which represents a given letter, is equal to one. Vertices of the unitary cube in $\mathbb{R}^4$ were tested as the alternative representation of the Polish letters. Seven encoded letters were put onto the perceptron input - the transformed one, the three preceding and the three following letters. The best perceptron achieved the accuracy of transformation equal to 96.4% on a testing set. In order to improve the efficiency, a modular system consisted of five neural networks was proposed. The individual perceptrons were specialized in the following subtasks: transformation of the digraph *rz*, transformation of nasal letters, transformation of the letter *n*, transformation of the letters *i,u,y*, according to the fact that in Polish language phonematic transformation of these letters is very similar, and the network which solved the problem of voicing and devocalization. The system was aided by nine rules because in Polish language translation of nine letters is independent of the context. This hybrid system achieved accuracy equal to 98.5%. The application of the voting committees allowed the system to achieve accuracy 99.2%. It is significantly better than accuracy of NETTALK system for phonematic transformation for English language proposed in [167]. The system consisted of a single perceptron. The version with one hidden layer achieved accuracy equal to 77% on testing set whereas the version with two hidden layers achieved asymptotic accuracy equal to 91%. Let us remember, however, that English is far more difficult to phonematic transformation than Polish.

Part II
Mathematical Foundations

Chapter 4
General Foundations

In this chapter general mathematical foundations, necessary for the presented studies, are specified. Most of them have basic character.

Let us start from the definition of a typical property. In general, the word *typical* refers to the property of the elements of a given set which is shared by all elements of a large subset of this set. There are various definitions about what it means that the set is large in such context. The one used in this monograph is a strong one.

Definition 4.1 A given property is said to be generic in a topological space if there exists an open and dense set in this space which has this property.

The dynamical systems on manifolds are the mathematical foundation for the analysis of the training process of perceptrons that is presented in this monograph. Let us recall some basic facts concerning manifolds - see, for instance, [148], Chap. 1 and [118], Chap. 2.

Let $\mathcal{M}$ be a subset of $\mathbb{R}^n$ with the induced topology on $\mathcal{M}$.

Definition 4.2 $\mathcal{M}$ is said to be a differentiable manifold of dimension m if for each $p \in \mathcal{M}$ there exists a neighbourhood $U_p \subset \mathcal{M}$ of p, an open set $V \in \mathbb{R}^m$ and a homeomorphism $f : U_p \to V$ such that the inverse homeomorphism $f^{-1} : V \to U_p \in \mathbb{R}^n$ is a C^∞ immersion. A pair (f, U_p) is called a chart at the point p. Furthermore, if $f^{-1} \in C^r$, then $\mathcal{M}$ is said to be a manifold of class r. (or C^r manifold).

Definition 4.3 Let $\mathcal{M}$ be a C^r manifold, $r \geq 1$. Let (U_1, f_1) be a chart at the point $p \in \mathcal{M}$. Let, furthermore, $\mathbf{v}_1$ be a vector which is an element of the vector space in which $f_1(U_1)$ is contained. It is said that two triples $(U_1, f_1, \mathbf{v}_1)$, and $(U_2, f_2, \mathbf{v}_2)$, specified at the same point p, are equivalent if the derivative of $(f_2 \circ f_1^{-1})$ at the point $f_1(p)$ maps $\mathbf{v}_1$ on $\mathbf{v}_2$. An equivalence class of such triples at the point p is called a tangent vector to $\mathcal{M}$ at p. The set of all tangent vectors to $\mathcal{M}$ at p is called a tangent space to $\mathcal{M}$ at p. It is denoted by $T_p(\mathcal{M})$.

© Springer International Publishing AG, part of Springer Nature 2019
A. Bielecki, *Models of Neurons and Perceptrons: Selected Problems and Challenges*, Studies in Computational Intelligence 770,
https://doi.org/10.1007/978-3-319-90140-4_4

Let us denote that each chart at the point $p \in \mathcal{M}$ determines a bijection of $T_p(\mathcal{M})$ onto a Banach space and, as a consequence, the structure of the topological vector space given by a chart is transported to $T_p(\mathcal{M})$. Let us also denote that the structure of the vector space $T_p(\mathcal{M})$ is independent of the choice of a chart.

Definition 4.4 Let $\mathcal{M}$ be an $n-$dimensional manifold. The tangent bundle of $\mathcal{M}$ is defined as

$$TM := \{(p, \mathbf{v}) \in \mathbb{R}^n \times \mathbb{R}^n : p \in \mathcal{M}, \mathbf{v} \in T_p M\} \subset \mathbb{R}^{2n}$$

with topology induced from $\mathbb{R}^n \times \mathbb{R}^n$.

It can be easily shown that TM is a differentiable $2n-$dimensional manifold and that the projection $\pi : TM \to \mathcal{M}$, given by the formula $\pi(p, \mathbf{v}) = p$, is continuous.

Two Whitney's theorems refer manifolds to Euclidean space. According to the first one each differentiable manifold can be regarded, in a natural way, as a submanifold of $\mathbb{R}^k$ - see [148], p. 9.

Theorem 4.5 (Whitney Theorem) *Let $\mathcal{M}$ be a differentiable $n-$dimensional manifold. Then, there exists a proper embedding $g : \mathcal{M} \to \mathbb{R}^{2n+1}$.*

Let us stress that, according to Whitney Theorem, not only $\mathcal{M}$ is a subset of $\mathbb{R}^{2n+1}$ but also the manifold together with the tangent bundle is embedded in $\mathbb{R}^{2n+1}$.

According to the second Whitney Theorem each differentiable manifold can be considered as a C^∞ manifold.

Theorem 4.6 (Whitney Theorem) *Let $\mathcal{M}$ be a C^r $n-$dimensional manifold, where $r \geq 1$. Then, there exists a C^r embedding $g : \mathcal{M} \to \mathbb{R}^{2n+1}$ such that $g(\mathcal{M})$ is a closed C^∞ submanifold of $\mathbb{R}^{2n+1}$.*

Flows on manifolds are generated by vector fields.

Definition 4.7 A C^r mapping

$$X : \mathcal{M} \to \mathbb{R}^n,$$

where $\mathcal{M}$ is an $n-$dimensional manifold, is called a C^r vector field on $\mathcal{M}$.

The definition means that a vector field transforms a point $p \in \mathcal{M}$ to $T_p M$.

Let $\langle \cdot, \cdot \rangle$ be a scalar product in $\mathbb{R}^j$. In each point p of the manifold $\mathcal{M}$ the space $\mathbb{R}^j$ generates scalar product $\langle \cdot, \cdot \rangle_p$ and, as a consequence, it generates the norm $|| \cdot ||_p$. Let $f : \mathcal{M} \to \mathbb{R}$ be a mapping of a class C^{r+1}. Then, for each $p \in \mathcal{M}$ there exists a unique vector $X(p) \in T_p M$ such that for each vector $\mathbf{v} \in T_p M$ the equality

$$df_p \mathbf{v} = \langle X(p), \mathbf{v} \rangle_p$$

is satisfied. In a such way on the manifold $\mathcal{M}$ a vector field f, called a gradient of a mapping f, has been defined. The mapping f is called a potential determined on $\mathcal{M}$. It can be shown that X is of a class C^r.

Let us move to algebraic foundations. In the analysis of properties of linear and weakly nonlinear perceptrons the following well known fact is used.

Theorem 4.8 *Let* $\mathbf{A}$ *be a real symmetric matrix. Then* $\mathbf{A}$ *has only real eigenvalues.*

Gram matrices are used in the analysis of dynamics of training process of the linear and weakly nonlinear perceptrons - see Chaps. 9 and 10. Gram matrices are symmetric and in this monograph only the real ones are considered. Let us recall the definition and some basic properties.

Definition 4.9 Let $\mathbf{v}_1, \ldots, \mathbf{v}_n$ be vectors in a space with a scalar product $\langle , \rangle$. The Gram matrix of these vectors, let us denote it as $\mathbf{G}(\mathbf{v}_1, \ldots, \mathbf{v}_n)$, consists of scalar products of these vectors which means that its elements are defined in the following way: $g_{ij} := \langle \mathbf{v}_i, \mathbf{v}_j \rangle$.

The definition means that the Gram matrix is generated by a finite sequence of vectors.

Corollary 4.10 *The vectors that generate Gram matrix are linearly independent if and only if the Gram matrix is nonsingular. If the generating vectors are elements of* $\mathbb{R}^k$, *then they are linearly independent if and only if* $\det \mathbf{G}(\mathbf{v}_1, \ldots, \mathbf{v}_n) > 0$ *and they are linearly dependent if and only if* $\det \mathbf{G}(\mathbf{v}_1, \ldots, \mathbf{v}_n) = 0$.

It turns that, under some natural assumptions, the linear independence of the family of vectors is a generic property.

Lemma 4.11 *Let* $\mathfrak{V}_n(m)$ *denotes the family of all* $m-$*elementary sets of vectors from* $\mathbb{R}^n$, *where* $m \leq n$. *Linear independency of vectors that belong to* $\mathfrak{V}_n(m)$ *is a generic property.*

The Lemma is a simple consequence of the fact that in the family of square real matrices the nonsingularity is a typical property.

The following property of Gram matrices are used in the analysis of the dynamical properties training process of linear perceptrons - see Chap. 9

Corollary 4.12 *For each sequence of* n *vectors that belong to* $\mathbb{R}^k$ *the following inequality holds*

$$\det \mathbf{G}(\mathbf{v}_1, \ldots, \mathbf{v}_m, \mathbf{v}_{m+1}, \ldots, \mathbf{v}_n) \leq \det \mathbf{G}(\mathbf{v}_1, \ldots, \mathbf{v}_m) \det \mathbf{G}(\mathbf{v}_{m+1}, \ldots, \mathbf{v}_n).$$

Chapter 5
Foundations of Dynamical Systems Theory

In this chapter the issues of dynamical systems theory, which is used for the further analysis, are presented. In the first section some basic facts are recalled. In the subsequent sections some advanced topics are considered. Then, the Euler method on a manifold is discussed. Then linear, weakly nonlinear and gradient dynamical systems are elaborated. In three last sections of this chapter topological conjugacy of cascades, pseudo-orbit tracing property and dynamical systems with control are presented. It should be mentioned that both topological conjugacy and shadowing property are the topics that are far from being worked out completely and a lot of problems in this field are open.

5.1 Preliminaries

Let us recall the most basic definitions and facts concerning dynamical systems theory.

Let X be a topological space and $(T, \oplus)$ be a topological group with a neutral element e.

Definition 5.1.1 A mapping $\Phi : X \times T \to X$ is a dynamical system if

1. $\Phi(x, e) = x$ for each $x \in X$;
2. $\Phi(\Phi(x, t_1), t_2) = \Phi(x, t_1 \oplus t_2)$ for each $x \in X$; and $t_1, t_2 \in T$;
3. Φ is continuous.

If $(T, \oplus) = (\mathbb{R}, +)$ then a dynamical system is called a flow; if $(T, \oplus) = (\mathbb{Z}, +)$ then a dynamical system is called a cascade.

© Springer International Publishing AG, part of Springer Nature 2019
A. Bielecki, *Models of Neurons and Perceptrons: Selected Problems and Challenges*, Studies in Computational Intelligence 770,
https://doi.org/10.1007/978-3-319-90140-4_5

Formally, a dynamical system should be denoted as (T, X, Φ). It is denoted, however, as (X, Φ) or simply as Φ if it does not lead to misunderstanding.

Let us assume that a dynamical system Φ is given.

Definition 5.1.2 The set

$$\mathrm{orb}_\Phi(x) := \{\Phi(x, t), t \in T\}$$

is called the <u>orbit</u> of the point x.

Definition 5.1.3 A point $p \in X$ such that $\Phi(p, t) = p$ for each $t \in T$ is called a <u>fixed point</u> of the system Φ.

Definition 5.1.4 A fixed point p of a dynamical system Φ, which is a flow or a cascade, is called <u>locally attracting</u> if there exists an open neighbourhood V_p of p such that for each $\overline{x \in V_p} \lim_{t \to \infty} \Phi(x, t) = p$, where $t \in \mathbb{R}$ in the case of a flow and $t \in \mathbb{Z}$ in the case of a cascade.

Definition 5.1.5 A fixed point p of a dynamical system Φ, which is a flow or a cascade, is called <u>locally repelling</u> if there exists an open neighbourhood V_p of p such that for each $\overline{x \in V_p} \lim_{t \to -\infty} \Phi(x, t) = p$, where $t \in \mathbb{R}$ in the case of a flow and $t \in \mathbb{Z}$ in the case of a cascade.

Definition 5.1.6 A point $p \in X$, which is not a fixed point, is called a <u>periodic point</u> if there exists $t_0 \in T$, $t_0 \neq e$ such that $\Phi(p, t_0) = p$.

In this monograph, dynamical systems on manifolds are used to analyse the dynamics of the learning (training) process of perceptrons. Let us recall the basic definitions - see [148], Chap. 1.

A specific dynamical system can be obtained in various ways. One of them is to give an algorithm of generation of its orbits.

A vector field generates orbits of a flow is such a way that a vector $\mathbf{v}_x$ at the point x is the velocity vector at the point x. Alternatively, it can be said that, for a given vector field X, the flow is generated by a differential equation

$$\frac{d\mathbf{x}}{dt} = X(\mathbf{x}) \tag{5.1}$$

with a given initial condition. Formally, for manifolds, it can be expressed in the following way - see [148], p. 11.

Proposition 5.1.7 *Let X be a C^r vector field on a compact manifold $\mathcal{M}$. Then, there exists a C^r mapping $\Phi : \mathcal{M} \times \mathbb{R} \to \mathcal{M}$ such that $\Phi(p, 0) = p$ and $\dfrac{d\Phi}{dt}(p, t) = X(\Phi(p, t)$ for each $p \in \mathcal{M}$ and $t \in \mathbb{R}$.*

The above proposition means that there exists a global flow Φ on $\mathcal{M}$ generated by X.

A bijection $F : \mathcal{M} \to \mathcal{M}$ generates orbits of a cascade by iterations, i.e. $\mathrm{orb}_F(x) := \{F^n(x), \ n \in \mathbb{Z}\}$, where F^{-1} denotes the mapping inverse to F and if n is a natural number then F^n denotes the nth superposition of the mapping F. When the cascades are considered on manifolds it is usually additionally assumed that the generating mapping F is a diffeomorphism.

Definition 5.1.8 A fixed point $p \in \mathcal{M}$ of the flow generated by a vector field X is called a <u>hyperbolic fixed point</u> if the derivative $dX_p : T_pM \to T_pM$ has no imaginary eigenvalues.

Definition 5.1.9 A fixed point $p \in \mathcal{M}$ of the cascade generated by a diffeomorphism F is called a <u>hyperbolic fixed point</u> if the derivative $dF_p : T_pM \to T_pM$ has no eigenvalues which have module equal to 1.

Definition 5.1.10 Let $W_\Phi^s(p)$ and $W_\Phi^u(p)$ denotes, respectively, <u>stable</u> and <u>unstable manifolds</u> of a fixed point p of a dynamical system Φ which is a flow or a cascade. These manifolds are the sets defined in the following way

$$W_\Phi^s(p) := \{x \in \mathcal{M}\} : \lim_{t \in \infty} \Phi(x, t) = p,$$

$$W_\Phi^u(p) := \{x \in \mathcal{M}\} : \lim_{t \in -\infty} \Phi(x, t) = p,$$

where $t \in \mathbb{R}$ in the case of a flow and $t \in \mathbb{Z}$ in the case of a cascade.

Definition 5.1.11 A hyperbolic fixed point which is neither attracting nor repelling is called a <u>saddle point</u>.

Definition 5.1.12 A dynamical system has not got saddle-saddle connections if the fact that a point x belongs to the stable manifold of a saddle point implies that x does not belong to the unstable manifold of any other saddle point.

So called limit sets play an important role in the description of properties of flows and cascades.

Definition 5.1.13 Let Φ be a flow or a cascade. The ω-limit set of a point $p \in \mathcal{M}$ is the set of those points $q \in \mathcal{M}$ for which there exists a sequence $t_n \to \infty$ such that $\Phi(p, t_n) \to q$. Similarly, the α-limit set of a point $p \in \mathcal{M}$ is the set of those points $q \in \mathcal{M}$ for which there exists a sequence $t_n \to -\infty$ such that $\Phi(p, t_n) \to q$.

Qualitative features of a dynamical system can be expressed by periodic-like behaviour of the regions of its phase portrait. The most strong examples of such behaviour are fixed points and periodic orbits. Nevertheless, more subtle possibilities are also taken into consideration. The following one plays an important role in a Morse-Smile dynamical systems widely considered in this book.

Definition 5.1.14 Let (X, Φ) be a dynamical system. A point $x \in X$ is said to be a non-wandering point if for any neighbourhood U_x of x there exists some $t_0 > 0$ such that $\Phi(U_x, t_0) \cap U_x \neq \emptyset$.

Clearly, the fixed points and periodic points are the non-wandering ones.

Let us define a Morse-Smile cascades and flows.

Definition 5.1.15 A cascade is called a Morse–Smale cascade if its non-wandering set is a finite union of periodic orbits and fixed points, each of which is hyperbolic one and whose stable and unstable manifolds are all transversal to each other. A mapping which generates a Morse–Smale cascade is called a Morse–Smale mapping.

Definition 5.1.16 A flow is called a Morse–Smale flow if its non-wandering set is a finite union of periodic orbits and fixed points, each of which is hyperbolic and whose stable and unstable manifolds are all transversal to each other, and there are no saddle-saddle connections. A vector field which generates a Morse–Smale flow is called a Morse–Smale vector field.

Definition 5.1.17 A dynamical system is called a Morse–Smale gradient-like dynamical system if it is a Morse–Smale system which has no periodic orbits.

The example of a Morse–Smale gradient-like flow on $\mathbb{R}^3$ is shown in Fig. 5.1. The following property of the presented flow should be noticed. Let us consider the ball which has the origin at the point A_1 and a sufficiently large radius. The trajectories of the flow cut transversally the sphere which is the boundary of the ball in a direction into the interior of the ball. Therefore, the flow can be completed after the compactification of $\mathbb{R}^3$ in such a way that it remains a Morse–Smale gradient-like flow with the additional repelling point. Such a complement plays a crucial role in the analysis of training process dynamics of nonlinear perceptrons - see Chap. 11.

If a flow $\Phi : \mathcal{M} \times T \to \mathcal{M}$ is given, then its time step can be fixed which gives, so called, discretization of the flow. In such a way a bijection $\Phi(\cdot, h) : \mathcal{M} \to \mathcal{M}$, usually denoted as Φ_h, is obtained. On the other hand, frequently, a numerical method is applied to a flow, especially if it is generated by a differential equation - see Eq. (5.1). If an operator of the used numerical method is a bijective mapping then it generates the cascade on $\mathcal{M}$. The question whether the cascade generated by a numerical method reflects the dynamics of the flow discretization is a classical topic in numerical dynamics. This is also one of the main topics of mathematical considerations presented in this monograph. This part of the mathematical research field is the basis of analysis of the perceptron training process.

Let us recall some basic facts that concern the stability theory.

Definition 5.1.18 The orbit $\mathrm{orb}_\Phi(x_0)$ of the flow Φ is called a stable orbit if for each $\varepsilon > 0$ and $t_0 \in \mathbb{R}$ there exists $\delta > 0$ such that the inequality $\varrho(\Phi(x_0, t_0), \Phi(x_1, t_0)) < \delta$ implies $\varrho(\Phi(x_0, t), \Phi(x_1, t)) < \varepsilon$ for each $t > t_0$.

Intuitively, the orbit is stable if orbits that start close to it at a time t_0 remains close to it for all times $t > t_0$.

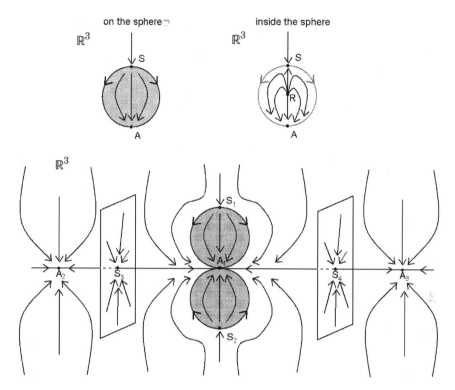

Fig. 5.1 The example of the phase portrait of a Morse–Smale gradient-like flow on $\mathbb{R}^3$. At the top, the flow on the ball, which is used for the construction of the whole system, is presented. The flow is shown of the sphere (left) and inside the ball (right). At the bottom, the phase portrait of the whole system is shown

Definition 5.1.19 The orbit $\mathrm{orb}_\Phi(x_0)$ of the flow Φ is called an asymptotic stable orbit if it is stable and for each $t_0 \in \mathbb{R}$ there exists $\Delta > 0$ such that the inequality $\varrho(\Phi(x_0, t_0), \Phi(x_1, t_0)) < \Delta$ implies $\lim_{t \to \infty} \varrho(\Phi(x_0, t), \Phi(x_1, t)) = 0$.

Definition 5.1.20 A flow is stable if its each orbit is stable. Similarly, a flow is asymptotically stable if each its orbit is asymptotically stable.

Definition 5.1.21 An asymptotically stable flow on the manifold $\mathcal{M}$ is said to be globally asymptotically stable if it has only one fixed point, let us say p, which is attracting and such that $W_\Phi^s(p) = \mathcal{M}$.

5.2 The Euler Method on a Riemannian Manifold

In this monograph the Euler method on compact Riemannian manifolds is considered. In such a case it is defined by using exponent mapping. Let us recall its definition.

The local exponent is a mapping $\exp_p : T_p M \supset U_p \to \mathcal{M}$ given by the formula

$$\exp_p(\mathbf{v}) = \gamma(1),$$

where U_p is a neighbourhood of the point p and γ is the geodesic line such that $\gamma(0) = \pi(p, \mathbf{v})$ and $\dot{\gamma}(0) = \mathbf{v}$. The dot denotes the derivative over the parameter by which the geodesic line is parametrized. The mapping $\pi : TM \to \mathcal{M}$ is the canonical projection i.e. it is given by the formula $\pi(p, \mathbf{v}) = p$.

Let $\mathcal{M}$ be a Riemanian C^j manifold, $j \geq 1$, embedded in $\mathbb{R}^{2k+1}$ - see Whithey Theorem. Let a flow on $\mathcal{M}$ be generated by the equation

$$\frac{d\mathbf{x}}{dt} = f(t, \mathbf{x}), \quad \mathbf{x}(0) = \mathbf{x}_o, \tag{5.2}$$

Let us assume that for every t the function $f(t, \cdot)$ is a Lipschitz mapping. Let the time step h be constant. The Euler method on $\mathcal{M}$ is defined in the following way

$$x_n = \exp_{x_{n-1}}(-h \cdot f(t_{n-1}, x_{n-1})), \tag{5.3}$$

where $f(t_{n-1}, x_{n-1})$ is a vector of the tangent space $T_{x_{n-1}} M$. If $\mathcal{M} = \mathbb{R}^k$ then the formula (5.3) has the form

$$x_n = x_{n-1} + h \cdot f(t_{n-1}, x_{n-1}), \tag{5.4}$$

where $t_n = t_{n-1} + h$.

Let the flow generated by the problem (5.2) be denoted as ϕ. Let, furthermore, ϕ_h and ψ_h be the cascades generated by the discretization $\phi(\cdot, h)$ and by the Euler method for the Eq. (5.2) respectively. For the initial points $x_0, \tilde{x}_0 \in \mathcal{M}$ the error $\tilde{e}_n(x_0, \tilde{x}_0, h)$ after n steps is defined as

$$\tilde{e}_n(x_0, \tilde{x}_0, h) := \varrho_R\left(\psi_h^n(x_0), \ \phi_h^n(\tilde{x}_0) \right), \tag{5.5}$$

where ϱ_R denotes a Riemannian metric on $\mathcal{M}$.

Let $T := [0, a]$, where $a = n \cdot h$. Let us define the maximal error after n steps

$$e_n(x_0, \tilde{x}_0, h) := \max_{k \leq n} \varrho_R\left(\psi_h^k(x_0), \ \phi_h^k(\tilde{x}_0) \right). \tag{5.6}$$

The error of the Euler method on a Riemannian manifold is estimated as follows - see [30].

Theorem 5.2.1

$$e_n(x_0, \tilde{x}_0, h) \leq e^{a \cdot L} \cdot e_0(x_0, \tilde{x}_0) + \frac{b}{L} \cdot \left(e^{a \cdot L} - 1\right) \cdot h, \tag{5.7}$$

where L and b are constants which depends on the starting point, the manifold properties and the length of the time interval.

Let us notice that if the manifold is compact then the constants L and b are global. Furthermore, in the case of a compact manifold, if the initial error $e_0(x_0, \tilde{x}_0)$ zeroes, which means $x_0 = \tilde{x}_0$, then the error (5.7) depends only on the time interval length a.

5.3 Linear Dynamical Systems

The linear flows are the simplest ones. Not only the analytical formula that defines their orbit can be obtained but also the analysis of their stability is simple. In this monograph the stability of perceptron training process, considered as dynamical system, is considered - see Chaps. 9–11. Let us recall some basic facts.

Definition 5.3.1 A flow Φ on $\mathbb{R}^n$ is called <u>linear homogenious</u> if it is generated by a differential equation

$$\frac{d\mathbf{x}}{dt} = \mathbf{A}\mathbf{x}$$

and <u>linear nonhomogeneous</u> if it is generated by a differential equation

$$\frac{d\mathbf{x}}{dt} = \mathbf{A}\mathbf{x} + \mathbf{b},$$

where $\mathbf{A}$ is an $n \times n$ matrix and $\mathbf{b}$ is an n-dimensional vector.

Lemma 5.3.1 *The flow generated by a linear nonhomogeneous equation $\frac{d\mathbf{x}}{dt} = \mathbf{A}\mathbf{x} + \mathbf{b}$ is asymptotically stable if and only if the flow generated by its homogeneous part $\frac{d\mathbf{X}}{dt} = \mathbf{A}\mathbf{x}$ is asymptotically stable.*

Lemma 5.3.2 *The flow Φ generated by a linear homogeneous equation is asymptotically stable if and only if $\lim_{t \to \infty} \Phi(\mathbf{x}, t) = \mathbf{0}$ for each $\mathbf{x}$.*

Corollary 5.3.3 *Asymptotically stable linear flow is globally asymptotically stable.*

Theorem 5.3.4 *The linear flow generated by the equation*

$$\frac{d\mathbf{x}}{dt} = \mathbf{A}\mathbf{x}$$

is asymptotically stable if and only if the real parts of all eigenvalues of the matrix $\mathbf{A}$ are negative.

Calculation or even approximation of eigenvalues for the matrices that have large size is usually troublesome. In the case of asymptotical stability of linear flows, however, the following Hurwitz criterion can be applied.

Let us recall that if a principal diagonal of a minor of a square matrix $\mathbf{A}$ consists of elements of the principal diagonal of this matrix, then the minor is called a principal minor.

Theorem 5.3.5 (Hurwitz criterion) *Let*

$$\frac{d\mathbf{x}}{dt} = \mathbf{A}\mathbf{x}$$

be a linear equation in $\mathbb{R}^n$, where $\mathbf{A}$ is a real matrix. The flow generated by this equation is asymptotically stable if and only if the following inequalities hold:

$\Delta_1 = -\mathbf{A}_1 > 0$

$\Delta_2 = -\mathbf{A}_1 \mathbf{A}_2 + \mathbf{A}_3 > 0$

$\Delta_3 = (-1)^3 \mathbf{A}_3 \Delta_2 > 0$

$\vdots$

$\Delta_n = (-1)^n \mathbf{A}_n \Delta_{n-1} > 0,$

where $\mathbf{A}_k$ is a sum of all principal minors of rank k of the matrix $\mathbf{A}$.

5.4   Weakly Nonlinear Dynamical Systems

The dynamics of linear dynamical systems is very regular. Nevertheless, most of the dynamical phenomena including, in particular, training process of perceptrons, are not linear and, as a consequence, they cannot be described by using the linear dynamical systems. Thus, the following problem emerges: does a class of nonlinear dynamical systems that are as regular as linear ones exist? It turns out that the answer is affirmative.

Let us consider a flow generated by a differential equation (5.1). The linear part can be separated and the equation can be written in the following form

$$\frac{d\mathbf{x}}{dt} = \mathbf{A}\mathbf{x} + g(\mathbf{x}). \tag{5.8}$$

If the nonlinear part g satisfies some conditions that concern limitations of its values and derivatives, then the flow generated by formula (5.8) is said to be weakly non-linear. The specific form of the aforementioned conditions depends on the context in which the problem is considered. The examples of such theoretical results which concern weakly nonlinear systems are given in Sect. 5.6. The application of the idea to perceptrons is discussed in Chap. 10.

5.5 Gradient Dynamical Systems

Gradient flows are the type of dynamical systems that are also very regular. Let us recall definition and some basic properties.

Definition 5.5.1 A dynamical system Φ on $\mathcal{M}$ is said to be gradient flow if it is generated by a differential equation of the following form:

$$\frac{d\mathbf{x}}{dt} = \text{grad} V(\mathbf{x}),$$

where the function $V : \mathcal{M} \to \mathbb{R}$ is called a potential.

Since $\text{grad} V(p) = 0$ if and only if $DV(p) = 0$, the point p is a fixed point of a gradient flow if and only if $DV(p) = 0$.

One of the most fundamental properties of gradient dynamical system is the fact that if a point is not a fixed point, then the value of the potential increases along its orbit. It can be expressed formally in the following form.

Theorem 5.5.2 *Let a point p does not be a fixed point of a gradient flow Φ. If $t_1 < t_2$, then $V(\Phi(p, t_1)) < V(\Phi(p, t_2))$.*

The following corollary is implied directly from the above theorem.

Corollary 5.5.3 *A gradient flow has no periodic orbit.*

Let us notice that the above corollary implies that each gradient system which is a Morse–Smale one is a Morse–Smale gradient-like system.

Corollary 5.5.4 *Both the limit sets of any orbit consists of fixed points.*

Corollary 5.5.5 *If a gradient flow on $\mathcal{M}$ has only finite number of fixed points then each limit set of any orbit consists of at most one point which is a fixed one. If the manifold $\mathcal{M}$ is compact then each limit set of any orbit consists of exactly one point which is a fixed one.*

The last corollary states that on a compact manifold each point belongs both to the stable manifold of an attracting or saddle fixed point and to the unstable manifold of a repelling or saddle point.

5.6 Topological Conjugacy

A flow, especially the one generated by a differential equation, can be analysed by using theorems that concern flows. On the other hand, cascades are usually hard to analyse because does not exist a mathematical tool for analysing discrete dynamical systems which is as strong as we have for continuous dynamical systems. Therefore, the crucial question occurs whether the cascade generated by a flow has dynamic properties similar to the ones that has the generating flow. Time $h-$map discretization - see Sect. 5.1 - is the simplest way to generate the cascade by using a flow. The orbits of Φ_h cascade are subsets of the orbits of Φ flow. Thus, in a way, Φ and Φ_h have the same dynamical properties although the first one is a flow whereas the second one is a cascade and, as a consequence, formally, they cannot be compared. The time discretization Φ_h cannot be implemented directly as a computational algorithm. The orbits can be calculated by using numerical scheme applied to the differential equation that generates the flow. The problem is whether the dynamics of the cascade Φ_h and the cascade, let us say Ψ_h, generated by the numerical method are the same or, at least, similar. Topological conjugacy is a formal tool that is used to study this problem.

Definition 5.6.1 Let (T, X_1, Φ_1) and (T, X_2, Φ_2) be dynamical systems. They are said to be topologically conjugate if there exists a homeomorphism $\alpha : X_1 \rightarrow X_2$ such that for each $x \in X_1$ and $t \in T$ the following is satisfied

$$\alpha(\Phi_1(x, t)) = \Phi_2(\alpha(x), t). \tag{5.9}$$

If the Eq. (5.9) is satisfied but the conjugating homeomorphism α is determined only on a neighbourhood U_p of a point $p \in X_1$, then the dynamical systems are said to be locally topologically conjugate at the point p.

In the case of cascades it is sufficient to satisfy Eq. (5.9) only for $n = 1$.

Corollary 5.6.2 *Let F_1 and F_2 be bijective mappings on X_1 and X_2 respectively. The cascades generated by F_1 and F_2 are topologically conjugate if and only if there exists a homeomorphism $\alpha : X_1 \rightarrow X_2$ such that*

$$\alpha \circ F_1 = F_2 \circ \alpha. \tag{5.10}$$

It should be stressed that, usually, the conjugating homeomorphism is not a diffeomorphism.

The conjugate dynamical systems have the same dynamics. In particular, orbits are transformed into orbits and their properties are preserved. This means that attracting, repelling and saddle stable points are images of attracting, repelling and saddle ones respectively and the dimension of stable and unstable manifolds of saddle points are preserved. The orbits are deformed in a homeomorphic way. Formally it can be expressed in the following way.

Theorem 5.6.3 *Let (X_1, Φ_1) and (X_2, Φ_2) be dynamical systems conjugate by a homeomorphism $\alpha : X_1 \rightarrow X_2$. Then:*

1. *For each $x \in X_1$, $\alpha(\mathrm{orb}_{\Phi_1}(x)) = \mathrm{orb}_{\Phi_2}(\alpha(x))$.*
2. *If $p \in X_1$ is a stable point of the system Φ_1, then $\alpha(p) \in X_2$ is a stable point of the same type of the system Φ_2.*

3. *If $x \in X_1$ is a periodic point of the system Φ_1, then $\alpha(x) \in X_2$ is a periodic point of the system Φ_2. Furthermore, the points have the same period.*
4. *If $Y \subset X_1$ is an invariant set of the system Φ_1, then $\alpha(Y) \subset X_2$ is an invariant set of the system Φ_2.*

As it has been already discussed, the linear dynamical systems have very regular dynamics because in the linear case the system has only one stable point. If the point is a hyperbolic one, then two systems in the same Euclidean space are conjugate if the stable manifolds of their stable point has the same dimension. This implies that the systems are conjugate if stable points are either both attracting or both repelling. Therefore the problem of topological conjugacy of two linear dynamical systems is trivial.

It turns out that near a hyperbolic fixed point the dynamics of a system is the same as the dynamics of a linear system which, in a way, corresponds to this hyperbolic fixed point. It happens both for cascades and flows and is formalized as Hartman Theorem for cascades and Grobman–Hartman Theorem for flows.

Theorem 5.6.4 (Hartman local linearization theorem) *Let Φ be a cascade generated by a diffeomorphism f on a manifold $\mathcal{M}$ and let $p \in \mathcal{M}$ be a hyperbolic fixed point of Φ. Let, furthermore, $A = Df_p : TM_p \to TM_p$. Then, there exist neighbourhoods $U_p \subset \mathcal{M}$ of the point p and $V_0 \subset TM_p$ of the origin in the tangent space, and a homeomorphism $\alpha : V_0 \to U_p$ such that the formula*

$$\alpha \circ A = f \circ \alpha$$

is satisfied on V_0.

Theorem 5.6.5 (Grobman–Hartman local linearization theorem) *Let Φ be the flow generated by a differential equation*

$$\frac{d\mathbf{x}}{dt} = f(\mathbf{x})$$

on a manifold $\mathcal{M}$ and let $p \in \mathcal{M}$ be a hyperbolic fixed point of Φ. Let, furthermore, $A = Df_p : TM_p \to TM_p$ and let Ψ be the linear flow generated by a linear differential equation

$$\frac{d\mathbf{x}}{dt} = \mathbf{A}\mathbf{x}.$$

Then, there exist neighbourhoods $U_p \subset \mathcal{M}$ of the point p and $V_0 \subset TM_p$ of the origin in the tangent space, and a homeomorphism $\alpha : V_0 \to U_p$ such that for each $\mathbf{x} \in U_p$

$$\alpha \circ \Psi = \Phi \circ \alpha$$

if only $\Phi(\mathbf{x}, t) \in U_p$.

The versions of Hartman and Grobman–Hartman theorems cited above say that dynamical systems have locally, near a hyperbolic fixed point, the same dynamics as the corresponding linear systems. The theorems, however, have also global versions.

Let us recall some basic facts and definitions. Let $F : \mathbb{R}^n \to \mathbb{R}^n$ be linear hyperbolic. This means, among others, that F has a spectrum $\sigma(F)$ which splits in two disjoint parts: $\sigma_s(F)$ inside the unit circle and $\sigma_u(F)$ outside the unit circle. The space $\mathbb{R}^n$ is a direct sum of two F–invariant subspaces, let us say E_s and E_u, where the spectra of $T_s := T \,|\, E_s$ and $T_u := T \,|\, E_u$ are $\sigma_s(F)$ and $\sigma_u(F)$ respectively. The number $a := \max\{\|T_s\|, \|T_u\|^{-1}\}$ is called the skewness of F with respect to the chosen norm.

As it has been mentioned in Sect. 5.4 linear systems preserve their regular dynamics under small perturbations. This property can be formalized in the following form.

Theorem 5.6.6 (Hartman global linearization theorem) *Let F be a linear hyperbolic automorphism of $\mathbb{R}^n$ with a skewness $a < 1$. Let $f : \mathbb{R}^n \to \mathbb{R}^n$ be a Lipschitzean mapping which has a Lipschitz constant $\kappa < \min\{1 - a, \|T_s^{-1}\|^{-1}\}$. Then F and $F + f$ are topologically conjugate.*

It turns out that discretization of the weakly nonlinear flow and its Euler method are topologically conjugate.

Theorem 5.6.7 (Fečkan Theorem) *Let $(\Phi, \mathbb{R}^n)$ be a flow generated by Eq. (5.8), where the linear operator A has no eigenvalues on the imaginary axis and $g \in C^1(\mathbb{R}^n, \mathbb{R}^n)$, $g(\mathbf{0}) = \mathbf{0}$, $\sup_{\mathbf{x} \in \mathbb{R}^n} |g(\mathbf{x})| < \infty$. Let, furthermore, $|Dg(\mathbf{x})| < b$, where $b > 0$ is sufficiently small. Let Φ_h be a discretization of the flow Φ and let Ψ_h be a cascade generated by the Euler method applied for the Eq. (5.8), which means that $\Psi_h(\mathbf{x}) = \mathbf{x} + h \cdot A\mathbf{x} + hg(\mathbf{x})$. Then, there exist a ball $B_r := \{\mathbf{x} : |\mathbf{x}| \le r\}$, a real number $h_0 > 0$ and a continuous mapping $\alpha : \mathbb{R}^n \times (0, h_0) \to \mathbb{R}^n$ such that for each $h \in (0, h_0)$ the mapping $\alpha_h : \mathbb{R}^n \to \mathbb{R}^n$ defined as $\alpha_h(\mathbf{x}) := \alpha(\mathbf{x}, h)$ is a homeomorphism and the following equation is hold*

$$\Phi_h \circ \alpha_h = \alpha_h \circ \Psi_h.$$

The constants in Fečkan theorem can be estimated.

Theorem 5.6.8 *Let Φ be a weakly nonlinear flow generated by Eq. (5.8), where the linear operator $\mathbf{A}$ has no eigenvalues on the imaginary axis. Let $\Psi_{\mathbf{A}_h}$ be a cascade generated by the Euler method applied to the linear part of the flow Φ i.e. $\Psi_{\mathbf{A}_h} := \mathbf{A}_h \mathbf{x}$, where $\mathbf{A}_h := Id + h\mathbf{A}$. Let us assume that $\mathbf{A}_h$ is hyperbolic and $0 < h < \|\mathbf{A}\|^{-1}$. Let Ψ_h be a cascade generated by the Euler method applied to Eq. (5.8) which means that $\Psi_h(\mathbf{x}) := \mathbf{A}_h(\mathbf{x}) + hg(\mathbf{x})$. Let, furthermore, $M_h := \max\{\|\mathbf{A}_h^s\|, \|(\mathbf{A}_h^u)^{-1}\|\}$. Under the notation of Theorem 5.6.7, if the below inequalities are satisfied*

$$h \cdot b < \frac{(1 - M_h)}{\|\mathbf{A}_h^{-1}\|} \tag{5.11}$$

$$h \cdot b(\|A\| + b) < \frac{(1 - M_h)}{\|A_h^{-1}\| \cdot \|e^{hA}\|} \tag{5.12}$$

then the conclusion of Theorem 5.6.7 holds.

From inequality (5.12) the following square inequality is obtained

$$hb^2 + h\|A\| \cdot b - \frac{1 - M_h}{\|A_h^{-1}\| \cdot \|e^{hA}\|} < 0. \tag{5.13}$$

Taking into account that b is positive, inequality (5.13) is satisfied for $b \in (0, b_1)$, where

$$b_1 = \frac{1}{2h} \left(-h\|A\| + \sqrt{h^2 \|A\|^2 + \frac{4h(1 - M_h)}{\|A_h^{-1}\| \cdot \|e^{hA}\|}} \right). \tag{5.14}$$

In a general case it is not an easy task to calculate the used matrix norms. Nevertheless, in the applications considered in this monograph - see Chap. 10 - the matrix A is real and symmetric. Therefore it is diagonalizable and it has only real eigenvalues. In such a case the matrix norms can be calculated easily and, as a consequence, the constants can be estimated effectively. Thus, provided that A is $n \times n$ matrix of a hyperbolic flow it is sufficient to consider the following form of it: $A = \text{Diag}(\lambda_1, \ldots, \lambda_k, \lambda_{k+1,\ldots,n})$, $\lambda_1 \geq \ldots \geq \lambda_k > 0 > \lambda_{k+1} \geq \ldots \lambda_n$. Let us denote $\lambda_{max} := \max\{|\lambda_1|, |\lambda_n|\}$. Thus, the matrix A can be written in the form

$$A = \begin{bmatrix} A_h^u & 0 \\ 0 & A_h^s \end{bmatrix},$$

where $A_h^u = \text{Diag}(1 + h\lambda_1, \ldots, 1 + h\lambda_k)$, $A_h^s = \text{Diag}(1 + h\lambda_{k+1}, \ldots, 1 + h\lambda_n)$. Furthermore, the matrix norms used in constant estimations in Fečkan Theorem - see Theorem 5.6.8 - have the simple forms:

$\|A\| = \lambda_{max}$,
$\|A_h\| = 1 + h\lambda_1$,
$\|A_h^{-1}\| = \frac{1}{1+h\lambda_n}$,
$\|(A_h^u)^{-1}\| = \frac{1}{1+h\lambda_k}$,
$\|A_h^s\| = 1 + h\lambda_{k+1}$,
$\|e^{hA}\| = e^{h\lambda_1}$.

The estimates (5.11) and (5.12) can be rewritten in the following forms

$$h \cdot b < (1 - M_h) \cdot (1 + h\lambda_n) \tag{5.15}$$

$$h \cdot b(\lambda_{max} + b) < \frac{(1 - M_h)}{(1 + h\lambda_n) \cdot e^{h\lambda_1}}. \tag{5.16}$$

To sum up, utilizing (5.14), the following estimation is obtained

$$0 < b < \min\{M_1, M_2\}, \tag{5.17}$$

where

$$M_1 := \frac{1}{h}(1 - M_h) \cdot (1 + h\lambda_n)$$

and

$$M_2 := \frac{1}{2h}\left(-h\lambda_{max} + \sqrt{h^2\lambda_{max}^2 + 4h(1 - M_h)(1 + h\lambda_n) \cdot e^{-h\lambda_1}}\right).$$

The problem of topological conjugacy between substantially nonlinear dynamical system is studied intensively. Some strong results have been obtained for Morse–Smale gradient-like dynamical systems. Let us recall the crucial results for this type of the dynamical systems on manifolds.

In the three theorems put forward below $\mathcal{M}$ is a finite-dimensional compact smooth Riemanian manifold without a boundary and

$$\Phi : \mathcal{M} \times \mathbb{R} \to \mathcal{M}$$

is a Morse–Smale gradient-like flow that is generated by a differential equation on the manifold $\mathcal{M}$

$$\frac{d\mathbf{x}}{dt} = F(x), \tag{5.18}$$

where F is a $\mathcal{C}^2$ vector field on $\mathcal{M}$. Furthermore, $\Phi_h : \mathcal{M} \to \mathcal{M}$ is the time-h-map of the system ϕ, i.e. $\phi_h(x) := \phi(x, h)$.

Theorem 5.6.9 *Let $\Psi_{h,p}$ denotes the diffeomorphism generated by a Runge–Kutta method of the stepsize h and order $p > 1$ which is applied to Eq. (5.18). Then for a sufficiently small $h > 0$ there exists a homeomorphism $\alpha_h : \mathcal{M} \to \mathcal{M}$ which conjugates the cascade generated by Φ_h and the cascade generated by the numerical operator $\Psi_{h,p}$ which means that the following formula holds:*

$$\Psi_{h,p} \circ \alpha_h = \alpha_h \circ \Phi_h. \tag{5.19}$$

Furthermore, $\lim_{h \to 0} \varrho(\alpha_h(x), x) = 0$.

The analogous theorem for the Runge–Kutta method of order $p = 1$, i.e. for the Euler method, is worked out only under very specific assumptions.

Theorem 5.6.10 *Let us assume that the manifold $\mathcal{M}$ is two-dimensional. Let Ψ_h denotes the diffeomorphism generated by the Euler method of the stepsize h which is applied to Eq. (5.18). Then for a sufficiently small h there exists a homeomorphism $\alpha_h : \mathcal{M} \to \mathcal{M}$ which conjugates the cascade generated by Φ_h and the cascade generated by a numerical operator Ψ_h which means that the following formula holds:*

$$\Psi_h \circ \alpha_h = \alpha_h \circ \Phi_h. \tag{5.20}$$

Furthermore, $\lim_{h\to 0} \varrho(\alpha_h(x), x) = 0$.

The topological conjugacy in all cases for Runge–Kutta methods applied for Morse–Smale dynamical systems on manifolds is worked out in the case of the iterative numerical operator.

Theorem 5.6.11 *Denote by $\psi_{h,k}$ the diffeomorphism generated by the Runge–Kutta method of the stepsize h and order k which is applied to Eq. (5.18). Let $T > 0$ be given. Then for a sufficiently large m and each $k \in \{1, 2, \ldots\}$ there exists a homeomorphism $\alpha_m : \mathcal{M} \to \mathcal{M}$ such that the following formula holds*

$$\psi^m_{\frac{T}{m},k} \circ \alpha_m = \alpha_m \circ \phi_T. \tag{5.21}$$

Furthermore, $\lim_{m\to\infty} \varrho(\alpha_m(x), x) = 0$.

5.7 Pseudo-orbit Tracing Property

The idea of topological conjugacy, that has been put forward in the previous section, is useful provided that the calculations are performed in exact arithmetic. This assumption, however, is not true. Therefore, in the analysis of dynamical properties of the training process of perceptrons the mathematical tool, which is appropriate for analysis of dynamical properties of the systems, whose orbits are calculated in approximate arithmetic, should be used. This leads us to the concept of shadowing property known also as the pseudo-orbit tracing property.

In this subsection some basic definitions and results concerning both the shadowing and the inverse shadowing property are recalled. They are applied then to analysis of the properties of perceptron training process - see Chap. 11.

Let us assume that $f : \mathcal{M} \to \mathcal{M}$ is a diffeomorphism, i.e. $f \in \text{Diff}(\mathcal{M})$. Let $\mathbb{Z}$ denotes the set of integer numbers.

Definition 5.7.1 Let Φ be a cascade generated by the mapping f. A sequence $\{y_n\}_{n\in\mathbb{Z}} \subset \mathcal{M}$ is called a δ-pseudo-orbit of Φ if, for each $n \in \mathbb{Z}$, the following inequality is satisfied

$$d(f(y_n), y_{n+1}) \le \delta.$$

Definition 5.7.2 The cascade Φ generated by f is shadowing, if for every $\varepsilon > 0$ there exists $\delta > 0$ such that any δ-pseudo-orbit $\{y_n\}_{n\in\mathbb{Z}}$ of the diffeomorphism f is ε-traced by the orbit of some point $x \in \mathcal{M}$, which means that for each $n \in \mathbb{Z}$ the following inequality is satisfied

$$d(y_n, f^n(x)) \le \varepsilon.$$

Let $M^{\mathbb{Z}}$ denotes the set of all sequences of elements which belong to M. Let us assume that the elements of the sequences are indexed by $\mathbb{Z}$. Let us recall the concept of δ-method.

Definition 5.7.3 A map $\mu_f : M \to M^{\mathbb{Z}}$ is called a δ-method of the diffeomorphism f, if the following conditions hold:

1. $\mu_f(y)_0 = y$, for all $y \in M$
2. $\mu_f(y)$ is a δ-pseudo-orbit of the mapping f.

Let us present the most general way to introduce the idea of the inverse shadowing. Let $\mathcal{T} = \mathcal{T}(f)$ denotes a collection of such δ-methods of f that for any $\delta > 0$ there exists a δ-method $\mu_f \in \mathcal{T}$. Such $\mathcal{T}$ will be called a class. The set of all δ-methods is then a class and it will be denoted by $\mathcal{T}_0$.

Definition 5.7.4 Let us distinguish four following classes of δ-methods.

1. The class $\mathcal{T}_c(f)$ of all continuous δ-methods, where the continuity of a δ-method μ is defined with respect to the product topology in $M^{\mathbb{Z}}$ - see [112].
2. Let $g : M \to M$ be onto map satisfying $D_\infty(f, g) \leq \delta$, where $D_\infty(f, g) := \sup_{x \in M} d(f(x), g(x))$. Define the mapping

$$\mu_f(y) = \text{orb}_g(y), \quad \text{for all } y \in M.$$

Then μ_f is a δ-method of f.
The class of all δ-methods of the above form, where the mapping g is a homeomorphism, is denoted as $\mathcal{T}_h(f)$.
3. Let $\chi_n : M \to M, n \in \mathbb{Z}$ be a family of maps such that $\chi_0 = id_M$ and for all n, $D_\infty(f \circ \chi_n, \chi_{n+1}) \leq \delta$ and let:

$$\mu_f(y) = \{\chi_n(y)\}_{n \in \mathbb{Z}}, \quad \text{for all } y \in M.$$

Then μ_f is a δ-method of f.
The class of all δ-methods of the above form, where each χ_n is a continuous map, is denoted as $\Theta_c(f)$.
4. Let $\chi_n : M \to M, n \in \mathbb{Z}$ be a family of maps such that for all n, $D_\infty(f, \chi_n) \leq \delta$ and let:

$$\mu_f(y) = \{y_n\}_{n \in \mathbb{Z}} \text{ such that } y_0 = y, \ y_{n+1} = \chi_n(y_n) \text{ for all } y \in M.$$

Then μ_f is a δ-method of f.
The class of all δ-methods of the above form, where each χ_n is a continuous map, is denoted as $\Theta_s(f)$.

Let $\mathcal{T}$ be a class of δ-methods.

Definition 5.7.5 It is said that the cascade generated by a mapping f has $\mathcal{T}-$ inverse shadowing property if for any $\varepsilon > 0$ there exists a positive δ such that for any orbit $\{x_n\}_{n \in \mathbb{Z}}$ and any δ-method $\mu_f \in \mathcal{T}$ there exists $y \in \mathcal{M}$ such that for all $n \in \mathbb{Z}$ the following inequality is satisfied

$$d(x_n, \mu_f(y)_n) < \varepsilon.$$

Definition 5.7.6 The cascade generated by a mapping f is $\mathcal{T} - \text{robust}$ (or $\mathcal{T}-$ bishadowing), if it is both shadowing and $\mathcal{T}-$ inverse shadowing.

Relations between various type of the introduced $\delta-$methods can be expressed in the form of the following corollary.

Corollary 5.7.7 *The $\delta-$methods satisfy the following relations.*

1. *If $\mathcal{T}_1(f) \subset \mathcal{T}_2(f)$ and f is $\mathcal{T}_2(f)$ inverse shadowing then it is $\mathcal{T}_1(f)$ inverse shadowing as well.*
2. *The following inclusions are satisfied:*

$$\mathcal{T}_h(f) \subset \Theta_c(f) \cap \Theta_s(f) \subset \Theta_c(f) \cup \Theta_s(f) \subset \mathcal{T}_0(f)$$

   and all the inclusions are proper. In particular, $\Theta_c(f)$ and $\Theta_s(f)$ do not include each other (see [152]).
3. *We have $\Theta_c(f) = \mathcal{T}_c(f)$. Namely, for a given δ-method $\mu \in \mathcal{T}_c(f)$ one can define maps χ_n as $\chi_n(y) = \mu(y)_n$ and this means that $\mathcal{T}_c(f) \subset \Theta_c(f)$. The other inclusion is obvious.*

Morse–Smale cascades are so regular that they have shadowing property. It can be expressed formally as the following theorem.

Theorem 5.7.8 *Let $\Theta := \Theta_c \cup \Theta_s$. Each Morse–Smale diffeomorphism is $\mathcal{T}-$ robust if $\mathcal{T} = \Theta$.*

Robustness is an invariant of topological conjugacy. In particular, the following theorem is satisfied:

Theorem 5.7.9 *Let f, $g : \mathcal{M} \to \mathcal{M}$ be topologically conjugate diffeomorphisms. For the class $\mathcal{T} = \Theta_c, \Theta_s$, $\mathcal{T}(f)$ robustness of f is equivalent to $\mathcal{T}(g)$ robustness of g.*

5.8 Dynamical Systems with Control

In control theory not only stability is studied but also controllability and observability of the system. In this section we recall only the properties that are used directly in the sequel.

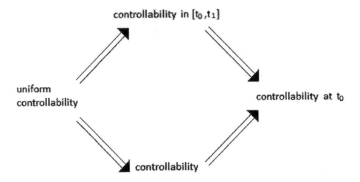

Fig. 5.2 Implications between various types of controllability

Let us consider the following linear problem

$$\frac{d\mathbf{x}}{dt} = \mathbf{A}\mathbf{x}(t) + \mathbf{B}\mathbf{v}(t), \tag{5.22}$$

$$\mathbf{y}(t) = \mathbf{C}\mathbf{x}(t) + \mathbf{D}\mathbf{v}(t), \tag{5.23}$$

in which the real matrices $\mathbf{A}$, $\mathbf{B}$, $\mathbf{C}$ and $\mathbf{D}$ do not depend on time, i.e. the problem (5.22)–(5.23) is time-invariant. Differential equation (5.22) describes the dynamics of the system state $\mathbf{x}(t) \in \mathbb{R}^n$ where $\mathbf{v} \in L^2_{loc}([t_0, \infty], \mathbb{R}^m)$ is the vector of accessible control, $L^2_{loc}([t_0, \infty], \mathbb{R}^m)$ denotes the space of locally $2-$integrable functions with values in $\mathbb{R}^m$. Algebraic equation (5.23) describes the output state. The matrices $\mathbf{A}$, $\mathbf{B}$, $\mathbf{C}$ and $\mathbf{D}$ have a dimension $n \times n$, $n \times m$, $p \times n$ and $p \times m$, respectively.

Definition 5.8.1 The dynamical system generated by Eq. (5.22) is said to be controllable in the time interval $[t_0, t_1]$ if for any vectors $\mathbf{x}(t_0) \in \mathbb{R}^n$ and $\mathbf{x}_1 \in \mathbb{R}^n$ there is a control $\mathbf{u} \in L^2_{loc}([t_0, \infty], \mathbb{R}^m)$ such that the trajectory $\mathbf{x}(t, \mathbf{x}(t_0), \mathbf{u})$ of the system satisfies the condition $\mathbf{x}(t_1, \mathbf{x}(t_0), \mathbf{u}) = \mathbf{x}_1$.

The system is controllable at time t_0 if there is $t_1 \in (t_0, \infty)$ such that the system is controllable in the time interval $[t_0, t_1]$.

The system is controllable if it is controllable for each $t_0 \in (-\infty, \infty)$.

The system is uniformly controllable if it is controllable for each time interval $[t_0, t_1]$.

It should be mentioned that the above definitions are usually put forward for the general case for which the matrices are time-dependent. In this general case the implications between the proposed types of controllability, shown in Fig. 5.2, can be directly derived from the definition.

For the time invariant case, however, the following lemma is satisfied.

Lemma 5.8.2 *For the time-invariant dynamical system (5.22) the notions of uniform controllability, controllability, controllability in the time interval $[t_0, t_1]$, and controllability at the point t_0 are pairwise equivalent.*

Since in the time-invariant case the characteristics of the equation do not change with time, it can be assumed, in this case, $t_0 = 0$, without loss of generality.

For the time-invariant dynamical systems the following simple criterium of their controllability exists.

Theorem 5.8.3 *The time-invariant dynamical system* (5.22) *is controllable if and only if*

$$\text{rank} \left[\mathbf{B} | \mathbf{AB} | ... | \mathbf{A}^{n-1} \mathbf{B} \right] = n,$$

where $\left[\mathbf{B} | \mathbf{AB} | ... | \mathbf{A}^{n-1} \mathbf{B} \right]$ denotes the block matrix with component matrices $\mathbf{B}$, $\mathbf{AB}, \ldots, \mathbf{A}^{n-1} \mathbf{B}$.

In practice, usually, some additional conditions are required, that concern the set of accessible controls. In a such context, a few additional types of controllability are proposed. Thus, let $M(V)$ denotes the set of vector measurable functions $\mathbf{v} :$ $(t_0, \infty) \to V \in \mathbb{R}^m$.

Definition 5.8.4 The time-invariant dynamical system (5.22) is said to be $V - $ controllable to the set $S \in \mathbb{R}^n$ from $\mathbf{x}_0 \in \mathbb{R}^n$ if for any initial state $\mathbf{x}(t_0) = \mathbf{x}_0$ there is an accessible control $\mathbf{v} \in M(V)$ such that there exists $t_1 > t_0$ such that $\mathbf{x}(t_1, \mathbf{x}(t_0), \mathbf{v}) \in S$. The system is locally $V-$controllable to the set $S \in \mathbb{R}^n$ if it is $V-$controllable to the set $S \in \mathbb{R}^n$ from each $\mathbf{x}_0 \in X$ and $S \subset X$. If $S = \{0\}$ then the system is locally $V-$controllable to zero.

Two following criteria of $V-$controllability can be specified.

Theorem 5.8.5 *Let zero belong to the interior of* V. *The system* (5.22) *is locally* V-*controllable to zero if and only if* rank $\left[\mathbf{B} | \mathbf{AB} | ... | \mathbf{A}^{n-1} \mathbf{B} \right] = n$.

Let $CH(V)$ denotes the convex hull of the set V.

Theorem 5.8.6 *The system* (5.22) *is locally V-controllable to zero if and only if it is locally CH(V)-controllable to zero.*

The idea of the system observability refers to the possibility of determination of the system states by using the values of the inputs and outputs of the system. In engineering applications, as well as in natural sciences, both the inputs and controls are given directly. In turn, the inner states of the system can be unobservable or at least difficult to observe. In such the cases observability of the system is a very useful property.

Definition 5.8.7 The system (5.22) and (5.23) is observable if its initial state $\mathbf{x}(0)$ can be determined on the basis of the control $\mathbf{v}(t)$ and the output $\mathbf{y}(t)$ over a finite-time interval.

The following criterion of observability is satisfied.

Theorem 5.8.8 *The system* (5.22) *and* (5.23) *is observable if and only if*

$$\text{rank} \left[\mathbf{C}^T | \mathbf{A}^T \mathbf{C}^T | ... | \left(\mathbf{A}^{n-1} \right)^T \mathbf{C}^T \right] = n.$$

Theorem 5.8.8 implies that the observability of a time invariant system (5.22) and (5.23) does not depend on the matrices $\mathbf{B}$ and $\mathbf{D}$ and, as a consequence, it is independent of control $\mathbf{v}$.

The set of all the systems that have the form (5.22) can be identified with the space $\mathcal{F}$ of pairs of the matrices because each system is determined unambiguously by the pair $(\mathbf{A}, \mathbf{B})$ of the $(n \times n)$−dimensional matrices $\mathbf{A}$ and $(n \times m)$−dimensional matrices $\mathbf{B}$.

It turns out that the controllability is a generic property of the time-invariant dynamical systems.

Theorem 5.8.9 *Let the set $\mathcal{F}$ be equipped with the topology induced by the metric ϱ defined as*

$$\varrho((\mathbf{A}, \mathbf{B}), (\mathbf{A}', \mathbf{B}')) := \|\mathbf{A} - \mathbf{A}'\|_1 + \|\mathbf{B} - \mathbf{B}'\|_2,$$

where $\| \cdot \|_1$ is a certain matrix norm equivalent to the Euclidean norm in the space of all $n \times n$ matrices whereas $\| \cdot \|_2$ is a matrix norm equivalent to the Euclidean norm in the space of all $n \times m$ matrices. Then, the set of time-invariant systems that are controllable is open and dense in $\mathcal{F}$ with respect to the introduced topology.

By Theorem 5.8.9, for almost all dynamical systems (5.22) there exists an open neighbourhood containing only controllable dynamical systems. Therefore, it is possible to define the controllability margin for a given dynamical system as the distance, according to the metric ϱ, between the present system and the nearest system that is not controllable.

5.9 Bibliographic Remarks

In this chapter both the basic foundations of the dynamical systems theory have been discussed and some advanced topics as well. There are many handbooks that concern the foundations of the dynamical systems - the books [99, 148, 183] can be put as examples. The stability of a dynamical systems as well as its linearization are also the basic topics. The first one is presented in detail in [61], whereas the second one - Hartman Theorem for cascades and Grobman–Hartman Theorem for flows can be found, for instance, in [148], Sect. 2.4. Some basic definitions and the properties concerning the controllability and observability of linear dynamical systems can be found, for example, in [109]. For the time-invariant dynamical systems the criteria of their controllability and observability are specified in [54], Sect. 5.4, [104], Sect. 1.9.5, [109], Sect. 1.10. The characteristics of various aspects of V-controllability can be found in [23, 24, 52, 93, 165] and [109], Sect. 1.9. The considerations that concern the controllability as a generic property of the time-invariant dynamical systems can be found in [55, 60], [109], Sect. 1.6, [122], Sect. 2.3, and [136].

The remaining part of the topics of this chapter concerns some advanced and specialized problems. Thus, discretization of flows in $\mathbb{R}^n$ is discussed in a series

Garay's papers [80–84] whereas the discretization on manifolds is the topic presented in [28, 31, 126–130].

The Fečkan Theorem has been published in [67].

The constants in Fečkan Theorem were estimated by Jabłoński - see [39, 101, 102].

Shadowing [46, 47, 111–113, 145, 146, 152], in particular the concept of δ-method was introduced by Kloeden and Ombach in [112]. The presented classes of δ-methods are discussed in [46, 152]. If $\mathcal{T} = \mathcal{T}_0$ then the definition 5.7.6 is the same as the concept of the inverse shadowing introduced and examined by Corless and Pilyugin in [57]. Actually, they did not use the notion of a δ-method there. In the same paper the authors showed that the definition was of a limited interest. In fact, they showed that any structurally stable diffeomorphism is not $\mathcal{T}_0$ inverse shadowing.

Corollary 5.7.7 can be found in [152].

Theorem 5.7.8 was proved in [46].

Part III
Mathematical Models of the Neuron

Chapter 6
Models of the Whole Neuron

A neuron, as it was discussed in Chap. 2, is a biological cell that has complex structure. Furthermore, numerous processes occur within it. Therefore, at the present level of scientific knowledge it is impossible to create any formal model that contains all the structural and dynamical aspects of the neuron. In such a situation two approaches can be applied: either a very simplified model of the neuron is created or there is created a model which describes only a part of a neuron structures or processes. The first group of the models is widely used as the basis for artificial neural networks. The second group of the models is frequently embodied as electronic circuits. Such an approach creates good perspectives for using the electronic circuits in future as the components of more holistic models of the neuron and, as the consequence, as the basis of artificial neural networks. In this chapter both groups of models are discussed. Connections with electronic circuits are presented as well.

Let us sum up the properties of the biological neuron described in Chap. 2. From the cybernetic point of view the neuron is a unit which processes signals. It has a few, let us say N, inputs - dendrites and one output - the axon. The neuron which has N inputs is called in the sequel an N−neuron. In general, there are two sorts of models of neurons: iterative and continuous. In the first one the input signals are put onto a neuron input iteratively, step by step. In the second one the input signal has continuous character. The iterative neurons and, what follows, the artificial neural networks based on them can be implemented by using both a software and digital electronic circuits. Continuous neurons and continuous artificial neural networks can be realized only by using analog circuits.

There are three types of models of the iterative neurons: deterministic, deterministic with memory and probabilistic.

In a deterministic neuron its inputs are weighted by the weights $w_1, \ldots, w_N$. The weighted inputs correspond to the dendrites, whereas the output corresponds to the axon. As it has been aforementioned, the input signals $x_1, \ldots, x_N$ are weighted which means that for each input the product $x_i \cdot w_i$ is calculated. Then, the products

© Springer International Publishing AG, part of Springer Nature 2019   59
A. Bielecki, *Models of Neurons and Perceptrons: Selected Problems and Challenges*, Studies in Computational Intelligence 770,
https://doi.org/10.1007/978-3-319-90140-4_6

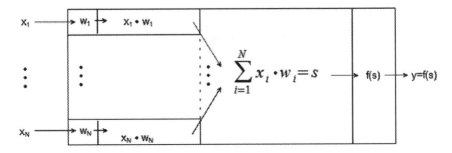

Fig. 6.1 The cybernetic schema of a neuron

are summed i.e. the so called total excitation $s = \sum_{i=1}^{N} x_i \cdot w_i$ of a neuron is calculated and, next, it is processed. The output signal y is the effect of this processing. The cybernetic schema of a such model of the neuron is presented in Fig. 6.1. The introduced model is a good starting point to create a mathematical model. The input signals and weights can be considered as vectors $\mathbf{x} = [x_1, \ldots, x_N] \in \mathbb{R}^N$ and $\mathbf{w} = [w_1, \ldots, w_N] \in \mathbb{R}^N$, respectively. Then, the total excitation s is a standard scalar product of vectors $\mathbf{x}$ and $\mathbf{w}$, so it can be put $s = \mathbf{x} \circ \mathbf{w}$ and, more general $s = \langle \mathbf{x}, \mathbf{w} \rangle$, where $\langle \cdot, \cdot \rangle$ denotes the inner product. The output signal y can be treated as a function of s, i.e. $y = f(s)$, $f : \mathbb{R} \rightarrow \mathbb{R}$, provided that the total excitation s is transformed in a deterministic way. The weights of a neuron are set, usually iteratively, during a training process. If the process is finished, then the neuron is called a trained one. Thus, a neuron which has N inputs, i.e. $\mathbf{x} \in \mathbb{R}^N$, is a family of functions indexed by a multi-index $\mathbf{w}$. Therefore a neuron can be identified with a mapping $F : \mathbb{R}^N \times \mathbb{R}^N \ni (\mathbf{x}, \mathbf{w}) \rightarrow F_{\mathbf{w}}(\mathbf{x}) \in \mathbb{R}$. Such a mapping embodied as an algorithm or a circuit is called an artificial neuron. An artificial neuron with fixed weights i.e. a mapping $F_{\mathbf{w}}(\cdot) : \mathbb{R}^N \rightarrow \mathbb{R}$ is called a trained neuron. The notation introduced above is regarded as the convention that will be obligatory in the rest of the monograph.

More general, a deterministic neuron can be regarded as a mapping

$$F : \mathbb{R}^n \times \Xi \ni (\mathbf{x}, \theta) \longmapsto F(\mathbf{x}, \theta) \in \mathbb{R},$$

where Ξ denotes the set of parameters indexing the family of the mappings. In the case of the aforementioned neurons based on the McCulloch's and Pitts's models [58] the weights of a neuron are, among others, the parameters of a neuron. The revival of the studies concerning artificial neural networks, which took place in the 1980s, resulted in, among others, using gradient methods for setting parameters of neural networks - the learning process of networks. Therefore, the differentiable functions has been widely used as the activation functions of a neuron since then. Linear neurons are the simplest ones of this type. Identity is the activation function of the linear neuron, i.e. $f(s) = s$. In this type of neurons $\Xi = \{\mathbf{w} \in \mathbb{R}^n\}$. The neuron with the identity function is the most general neuron that realizes the linear mapping which can be formally expressed as follows.

Lemma 6.1 *Let* $A : \mathbb{R}^N \rightarrow \mathbb{R}$ *be a linear operator. There exists a trained linear* N-*neuron that realizes the operator* A.

Proof Let a vector basis be given. In this basis the operator A can be expressed as a one-row matrix $[a_1, \ldots, a_N]$. If the neuron weights are set as $w_1 = a_1, \ldots, w_N = a_N$, then for each $\mathbf{x} = [x_1, \ldots, x_N]^T \in \mathbb{R}^N$ the value of the neuron output $y(\mathbf{x})$ is equal to $\mathbf{A}\mathbf{x}$, where $\mathbf{A}$ is the matrix which, in a given basis, corresponds to the operator A.

□

Linear neurons and the artificial neural networks consisted of them, called linear neural networks, have very limited abilities. Therefore, nonlinear activation functions are widely used. The logistic function is contemporary the most common one:

$$f_\beta(s) = \frac{1}{1 + \exp(-\beta s)} \tag{6.1}$$

In the models based on this type of function $\Xi = \{\mathbf{w} \in \mathbb{R}^n, \beta \in \mathbb{R}\}$. Historically, however, binary neurons were used as the first ones. They process signals by using the Heaviside function

$$f(s) = \begin{cases} 0 \text{ if } s \leq p_0, \\ 1 \text{ if } s > p_0. \end{cases} \tag{6.2}$$

In the case of a binary n-neuron $\Xi = \{\mathbf{w} \in \mathbb{R}^n, p_0 \in \mathbb{R}\}$. They were studied intensively in the 1960s in the context of logic. The obvious question is whether the artificial neural network can realize logical calculi. In such a context the problem whether each two-argument logical operator can be realized by a single neuron is the most basic one. It turned out that fourteen of the all sixteen two-argument binary operators could be realized by the binary neuron - see the following lemma.

Lemma 6.2 *Each of the two-argument binary logical operator apart from the equivalence and XOR operators can be realized by a binary 2-neuron.*

Proof Let us consider the problem from geometric point of view. Each binary logical operator is a binary function defined on the four-component set $X = \{(x_1, x_2) : x_1, x_2 \in \{0, 1\}\}$. On the other hand, the border line between 1 and 0 values of the output for the trained binary 2-neuron is given by the equation $x_1 \cdot w_1 + x_2 \cdot w_2 - p_0 = 0$ which is the equation of the straight line in $\mathbb{R}^2$ if w_1, w_2 and p_0 are set. Thus, the problem is reduced to the question: For which binary two-argument operators the points for which the operator value is equal to 1 (true value) can be separated by a straight line from those points for which the operator value is equal to 0 (false value). It is obvious that it is possible for each of two constant operators and for those ones for which the value of the operator for exactly three points is the same. In the last case one vertex of the unit square should be separated from the three others. For the operators that have the same value on exactly two points, separation by a straight line is possible only for these cases for which the points with the same operator value

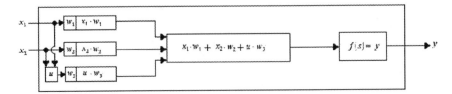

Fig. 6.2 The cybernetic schema of the multiplex neuron which realizes the XOR operator, $u = x_1 \cdot x_2$

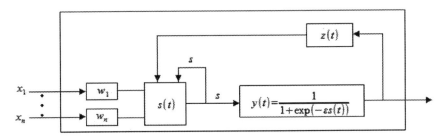

Fig. 6.3 The cybernetic schema of the continuous-time neuron proposed in [56]

lie on the edge of the unit square. If they lie on a diagonal, as it is in the case of the equivalence and XOR operators, then the separation by a single straight line is impossible.

$\square$

As it has been aforementioned in Chap. 2 a neuron is a multiplex module which means that the signals from dendrites are processed also before they reach the axon. It has been modelled in such a way that the processed input signals are put at the additional weighted neuron input - see Fig. 6.2 for a 2-neuron. Let such type of an artificial neuron be called a multiplex neuron. Let us introduce a multiplex multiplying 2-neuron i.e. a neuron that has one additional weighted input on which the signal $u(x_1, x_2) = x_1 \cdot x_2$ is put (Fig. 6.3).

Lemma 6.3 *Each of the two-argument binary logical operator can be realized by a multiplex multiplying binary 2-neuron.*

Proof The border line between 1 and 0 values is of the form: $x_1 \cdot w_1 + x_2 \cdot w_2 + u \cdot w_3 - p_0 = 0$, where $u(x_1, x_2) = x_1 \cdot x_2$. A binary 2-neuron is obtained by putting $w_3 = 0$. Therefore, a multiplex multiplying binary 2-neuron has not less computational abilities than a binary 2-neuron. Thus, it is sufficient to show that it can realize two operators that are nonseparable linearly. It can be easy verified that the multiplex neuron with $w_1 = w_2 = 2$, $w_3 = -4$ and $p_0 = 1$ realizes XOR operator whereas the neuron which realizes the equivalence operator can be obtained by putting, for instance, $w_1 = w_2 = -2$, $w_3 = 4$ and $p_0 = -1$.

$\square$

Radial neurons are another class of models of neurons. In this type of deterministic neurons the function, that is realized by a neuron, has a radial-based form which means that their values are changed radially around a given centre $\mathbf{c} \in \mathbb{R}^n$. The functions

$$f(r) = \exp\left(-\frac{r^2}{2\sigma^2}\right)$$

and

$$f(r) = g(r^2 - \sigma^2),$$

where $r = ||\mathbf{x} - \mathbf{c}||$, are most often used as radial-based ones. The set of the parameters of a radial neuron has the form $\Xi = \{\mathbf{c} \in \mathbb{R}^n, \sigma \in \mathbb{R}\}$.

Such neurons in which the output signal is determined not only by the current excitation but also by the value of the previous neuron excitation or by the previous state of its output are the generalization of the above-mentioned models

$$\begin{aligned} y(0) &= y(s(0)) \\ y(k+1) &= f(s(k+1), s(k), y(k)). \end{aligned} \tag{6.3}$$

A neuron with a hysteresis is the intensively studied type of such neurons. The simplest example of activation function has the following form

$$y(k+1) = \begin{cases} 1 & \text{dla } s(k+1) > p_g \\ y(k) & \text{dla } p_d \leq s(k+1) \leq p_g \\ 0 & \text{dla } s(k+1) < p_d, \end{cases} \tag{6.4}$$

where p_g, p_d denote upper and lower threshold respectively. The set of the parameters of a neuron with hysteresis has the form $\Xi = \{\mathbf{w} \in \mathbb{R}^n, p_g, p_d \in \mathbb{R}\}$.

A self-exciting neuron is another simple example of a neuron determined not only by its current state:

$$\begin{aligned} s(k+1) &= w \cdot y(s(k)) + x(n+1) + p_0, \\ y(k) &= \frac{1}{1+\exp(-s(k))}. \end{aligned} \tag{6.5}$$

The set of the neuron parameters is of the form $\Xi = \{w \in \mathbb{R}, p_0 \in \mathbb{R}.\}$. A neuron with the memory, in which all previous values of the output signals affect the current output value, is another possible generalization. It acts according to the following schema:

$$y(k+1) = f_\beta\left(\sum_{j=1}^n w_j x_j - \sum_{r=0}^k \eta^r y(k-r)\right), \tag{6.6}$$

where f is an activation function that depends on the set of parameters β (usually $\beta \in \mathbb{R}$,) and $\eta \in (0, 1)$ is a dumping factor. As it is implied by formula (6.6), the more distant in time the previous state is the less influence on the current state it has. It this model $\Xi = \{\mathbf{w} \in \mathbb{R}^n, \eta \in (0, 1), \beta\}$.

All above-mentioned models are the discrete ones which means that they can be implemented directly on computers. The continuous models, described usually by differential equations, are the second basic group of deterministic neurons. Such models can be implemented directly only by using analog electronic circuits. On computers they can be implemented only after discretization. In such a case the basic question is whether the crucial properties of the model are preserved under discretization - see Sects. 5.6 and 5.7. There are several continuous models of a neuron. One of the most general one was proposed in [56]. The model is based on ordinary differential equations:

$$
\begin{aligned}
\frac{ds(t)}{dt} &= k \cdot s(t) + \alpha \left(\sum_{n=1}^{N} w_n \cdot x_n(t) + p_0 \right) - z(t)(y(t) - p_1), \\
\frac{dz(t)}{dt} &= -\beta z(t), \\
y(t) &= \frac{1}{1+\exp(-\varepsilon s(t))},
\end{aligned}
\tag{6.7}
$$

where the nonnegative term $z : \mathbb{R} \to [0, \infty)$ is related to inhibitory self-feedback with a bias p_1. The above model is a generalization of the historically first model based on differential equations proposed by Hopfield [95]. The only difference between these two models is that a nonlinear term $-z(t)(y(t) - p_1)$ is added. As a result, the dynamics of the neuron depends strongly on z.

In the deterministic neurons, if the neuron parameters are set, then the value of the input signal determines unambiguously the value of the output signal. Probabilistic neurons are another group of models in which the value of the output signal is drawn according to a given probabilistic distribution. The following probabilistic binary neuron can be presented as a simple model of this type

$$
\begin{aligned}
p(y = +1) &= \frac{1}{1+\exp(-2\beta s)} \\
p(y = -1) &= \frac{1}{1+\exp(+2\beta s)},
\end{aligned}
\tag{6.8}
$$

where $s = \mathbf{x} \circ \mathbf{w}$, $p(y = +1)$ is probability that the value of the output signal is equal to 1. In the probabilistic neurons described by formulae (6.8) the set of the parameters of the neuron is of the form $\Xi = \{\mathbf{w} \in \mathbb{R}^n, \beta \in \mathbb{R}, P\}$, where P is a probability distribution.

In the light of neurophysiological knowledge, the models of the whole neuron are simplified to such an extent that they do not reflect, even approximately, the character of signal processing in the biological neuron. For instance, on the basis of the analysis of the functional properties of the neuron done by Waxman four fragments of the signal processing in the neuron was distinguished: dendritic region, the body of the cell, the axonal segment and the synapse. Furthermore, the axon can modulate spatial and temporal relations between signals acting as an active filter. Nevertheless, artificial neural networks built from even such simple units both model some neurophysiological phenomena and can be used as effective systems of artificial intelligence. Furthermore, such artificial neural networks generate interesting mathematical problems. Some aspects of them are the topics of Chaps. 8–11 and Appendix 13 of this monograph.

6.1 Bibliographic Remarks

The introduced model is based on the idea presented in [58].

Radial-based neurons were introduced in 1988 - [53, 140].

Neurons with hysteresis were proposed in 1991 - [178].

A self-exciting neuron is described in [150].

The neuron with the memory is introduced in [3].

The Waxman model was introduced in [180] and it is discussed in [166], Chap. 5.

The discussion concerning the axon functionality, including modulating spatial and temporal relations between signals is presented in [173].

Probabilistic binary neuron is presented [90], Sect. 5.6.

Firing neurons, omitted in this monograph as the subject of discussion, are described, for instance, in [132].

Electronic circuits are the widely used models of functional aspects of both whole neurons and their parts since the early 1960s when an electronic model of the whole neuron was proposed in [86], see also [176], Chap. 4.

Chapter 7
Models of Parts of the Neuron

As it has been mentioned in the previous section, the models of the whole neuron are too simplified to reflect all crucial aspects of the signal processing which is performed by the nervous system. Therefore, models of parts of the neuron are created. Both mathematical models and electronic circuits are used for modelling the parts of neural cells. These two approaches refer to each other - if a circuit model is given, then the ordinary differential equation that describes the dynamics of potential or current in the circuit can be formulated. On the other hand, mathematical formulae can be often realized by using the circuit that acts according to the dynamics described by these formulae. In this section both approaches are exploited and the relationships between them are discussed.

Four regions of the neuron have been distinguished and, as the consequence, described in the subsequence subsections: the dendritic region, the axon, the presynaptic bouton and the synapse.

It should be stressed that, currently, the computational power of computers is too weak to compose the model of the whole neuron by using models of its parts. Therefore the results described in this section have not been applied yet in the context of artificial neural networks. The observed progress in hardware development may enable the researchers to implement artificial neuron and, a consequence, artificial neural networks built from neurons based on prototypes combined from the models of parts of the neuron.

7.1 Model of Dendritic Conduction

The fact that the cellular membrane separates the ions on its internal and external surface is the starting point for many models of conduction in the neuron. On the other hand, the capacitor is the element of electric circuits which separates charge.

© Springer International Publishing AG, part of Springer Nature 2019
A. Bielecki, *Models of Neurons and Perceptrons: Selected Problems and Challenges*, Studies in Computational Intelligence 770,
https://doi.org/10.1007/978-3-319-90140-4_7

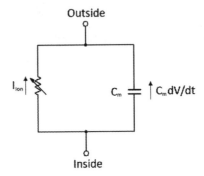

Fig. 7.1 The electronic model of a segment of the cellular membrane

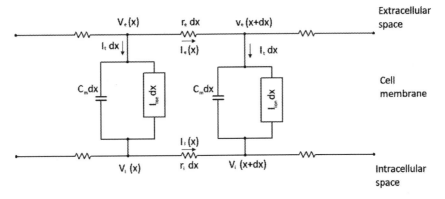

Fig. 7.2 The electronic model of the cellular membrane considered as discretized cable - see [106], Sect. 8.1

Thus, the simplest model of a segment of the cellular membrane can be built by using a simple circuit with the capacitor and the resistor - see Fig. 7.1. This basic idea appears, in various variants, in the sequel in this chapter as the basis of the model of the membrane conductance.

Utilizing Coulomb and Kirchoff laws, the equation that describes the dynamics of potential in the circuit presented in Fig. 7.1 can be easily obtained

$$c\frac{dV}{dt} + I_{ion} = 0, \qquad (7.1)$$

where $V = V_{inside} - V_{outside}$.

The series connection of the circuits that model a segment of the cellular membrane leads to the model of a long fragment of the cellular membrane - see Fig. 7.2. It can be shown that, after introducing the dimensionless variables, the dynamics of voltage changes is described by, so called, the cable equation

$$\frac{\partial V}{\partial \tau} = \frac{\partial^2 V}{\partial X^2} + f(V, \tau), \tag{7.2}$$

where τ is a dimensionless time and X is a dimensionless length. Normal electrical activity of neuronal dendrites has passive character which means that the linear form of the Eq. (7.2) is a good approximation

$$\frac{\partial V}{\partial \tau} = \frac{\partial^2 V}{\partial X^2} - V. \tag{7.3}$$

Equation (7.3) needs to be complemented by some initial and boundary conditions. It is, usually, assumed that initially the dendrite is in the resting state which means $V(X, 0) = 0$.

Various boundary conditions can be specified in dependence on the assumptions that concern the conditions at the boundaries of the dendrite. For instance, if the voltage is fixed at the boundaries, then $V(X_b, \tau) = V_b$, where X_b denotes the end of the dendrite. It is assumed that if a current, let us say $I(\tau)$, is injected at one end of the cable, then the boundary condition is of the form $\frac{\partial V(X_b, \tau)}{\partial X} = \alpha I(\tau)$.

7.2 Model of Axonal Transport

Processing of the impulse in the axon takes place in the cellular membrane. The Hodgkin–Huxley system of ordinary nonlinear differential equations is the first and the most classical model of the process which was confirmed experimentally and is regarded as the classical basis of neurodynamics. Since the model is commonly known - see the bibliographic remarks - it is not discussed in this book. High computational complexity of the model was the reason for looking for simpler models. The FitzHugh–Naguno proposal is one of them.

The FitzHugh–Naguno proposal is a simplified model of the cellular membrane. In their approach three aspects of the membrane are modelled by three components of the electronic circuit and the components are connected parallelly - see Fig. 7.3. The first component, which consists of the capacitor, models the membrane capacitance. The second one, which consists of the resistor, coil and battery connected in series, represents the recovery current. The third component, the nonlinear one, models the fast current. Such an approach leads to the following equations that describe the system dynamics:

$$C_m \frac{dV}{dt} + F(V) + i = -I_0, \tag{7.4}$$

$$L \frac{di}{dt} + Ri = V - V_0. \tag{7.5}$$

Fig. 7.3 The electronic
circuit functionally
equivalent to the FitzHugh
model

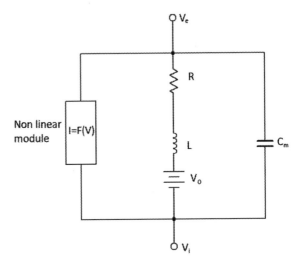

In the above equations i is the current in the line including the coil, I_0 is the external current, $V = V_i - V_e$ is the membrane potential, V_0 is the battery potential - see Fig. 7.3. The nonlinear module is described by the function $F(V) = I$.

After introducing dimensionless constants: $\varepsilon = \frac{R_1^2 C_m}{L}$, $\gamma = \frac{R}{R_1}$, $w_0 = \frac{R_1 I_0}{V_1}$, $v_0 = \frac{V_0}{V_1}$ and dimensionless variables: $\tau = \frac{Lt}{R_1}$, $v = \frac{V}{V_1}$, $w = \frac{R_1 i}{V_1}$, $f(v) = -\frac{R_1 F(V_1 v)}{V_1}$, the system (7.4)–(7.5) takes the form:

$$\varepsilon \frac{dv}{d\tau} = f(v) - w - w_0, \tag{7.6}$$

$$\frac{dw}{d\tau} = v - \gamma w - v_0. \tag{7.7}$$

The variable v is the fast one whereas w represents the slow variable. Since the Hodgkin–Huxley model is confirmed experimentally, the FitzHugh–Naguno model is required to have similar properties to the Hodgkin–Huxley model. In order to ensure this, the nonlinear component f should be similar to cubic polynomial which has one zero point at the origin of the coordinate system and two other positive. Thus, the polynomial of the form $f(v) = av(v - \alpha)(1 - v)$, where $0 < \alpha < 1$, is the classic choice. Piecewise linear model, which consists of three linear segments, is another possibility that is used commonly. It should be also mentioned that the electronic circuit presented in Fig. 7.3 is not the only analog device that models the dynamics described by Eqs. (7.6)–(7.7) - the more complex circuit in which operational amplifiers are used, is another possibility. The system can be analyzed by using phase-plane technique which is caused by the fact that the FitzHugh approach leads to the two-variable dynamical system. It was shown that if the model is considered as a family of dynamical systems indexed by the parameter ε then a bifurcation occurs.

To sum up, this significantly simplified model of the dynamics of signal processing by the axon demonstrates that this processing is complex, in particular nonlinear, and the character of the dynamics can change.

7.3 Models of Transport in the Presynaptic Bouton

Signal transmission between neurons consists in either transport of chemical substances from the presynaptic neuron to the postsynaptic one or direct transfer of ions between the neurons. In the first case the phenomena that take place in a presynaptic bouton are the first stage of the process. They are modelled by using either ordinary or partial differential equation. In the three subsequent subsections both types of the models are presented.

7.3.1 The A-G Model of Fast Transport Based on ODEs

The transport phenomena in biological units, including transport of various substances at subcellular level, are controlled precisely. In the commonly used compartment models that are based on ODEs the transport phenomenon is described as flows via channels between reservoirs (pools) in which the medium is stored. Aristizabal and Glavinovic created the compartment model of vesicular storage and release (A-G model for abbreviation). In this model the vesicles can be stored in three pools - the immediately available one, the small one and the large one, in the dependence on the proximity to the membrane of the cell and the degree of release competence. The vesicles in the immediately available pool, about 20% of all the vesicles, correspond to the docked vesicles - see Chap. 2. They are released to the synaptic cleft when the action potential arrives. Then, the pool is replenished by the flow from the small pool that, in turn, is replenished by vesicles from the large one. The whole process is described by the system of ordinary differential equation in which the density of vesicles is the unknown quantity.

The use of analog electronic circuits as models of biological phenomena are standard approach in biological modelling, including subcellular processes. It should be stressed, however, that electronic models are not an alternative to the ODE models but they are a fast and efficient way of physical realization of the processes described by differential equations. Such realization is useful in order to represent the mechanisms occurring both in biological and artificial structures, for instance, in robotics or in hardware realization of artificial neural networks. On the other hand, the electronic model can sometimes be obtained in easier way than a differential one. In such a case, a mathematical model based on differential equations can be obtained easily as a description of a circuit dynamics.

In general, in modelling of transport phenomena, the capacitor corresponds to a reservoir, voltage corresponds to the factor which causes the transport and the

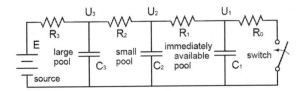

Fig. 7.4 The scheme of the electric circuit which corresponds to A-G model - see [5]. Capacitors represent the pools of vesicles, resistors - the resistance forces during diffusion between the pools. The source corresponds to the synthesis of neurotransmitter and the switch models the process of neurotransmitter release

resistor models the resistance forces, for instance the friction. Thus, let C_i, $i \in \{1, 2, 3\}$, denotes the ith pool in which vesicles are stored and the pool is modelled by a capacitor - see Fig. 7.4. Let, furthermore, U_i denotes the unknown vesicles density modelled by voltage and E denotes the speed of production of vesicles by the source that is situated in the large pool marked by subscript 1. Moreover, $\frac{1}{R_i C_j}$ denotes replenishment rates of various pools. The small and large pools are marked by subscripts 2 and 3 respectively and subscript 0 marks the synaptic release channel. When the action potential arrives, then the channel is open and the dynamics of the vesicular transport is described by the following differential model:

$$\frac{dU_1}{dt} = -\left(\frac{1}{R_1 C_1} + \frac{1}{R_0 C_1}\right) U_1 + \frac{1}{R_1 C_1} U_2, \tag{7.8}$$

$$\frac{dU_2}{dt} = \frac{1}{R_1 C_2} U_1 - \left(\frac{1}{R_2 C_2} + \frac{1}{R_1 C_2}\right) U_2 + \frac{1}{R_2 C_2} U_3, \tag{7.9}$$

$$\frac{dU_3}{dt} = \frac{1}{R_2 C_3} U_2 - \left(\frac{1}{R_3 C_3} + \frac{1}{R_2 C_3}\right) U_3 + \frac{1}{R_3 C_3} E. \tag{7.10}$$

The above system of differential equations describes the dynamics of electric voltage in the circuit which consists of three capacitors and a source connected in parallel. Furthermore, the circuit includes four resistors and a switch - see Fig. 7.4. The closed switch corresponds to the release of neurotransmitter during stimulation by the action potential. The open switch prevent the capacitor C_1 from discharge which corresponds to $R_0 \to \infty$. Thus, the Eq. (7.8) takes the following form:

$$\frac{dU_1}{dt} = -\frac{1}{R_1 C_1} U_1 + \frac{1}{R_1 C_1} U_2. \tag{7.11}$$

To sum up, if the switch is closed then the dynamics of the electronic circuit is represented by Eqs. (7.8)–(7.10). Otherwise it is represented by Eqs. (7.9)–(7.11). In the circuit E denotes the voltage of the source, U_i is the potential across the ith capacitor, C_i denotes the capacitance of the ith capacitor and R_i is resistance of the ith loop.

7.3.2 The Model of Fast Synaptic Transport Based on PDEs

In this subsection the model of fast transport of neurotransmitters, based on partial differential equations (PDEs, for abbreviation), is presented. In the model the following parameters are used.

(i) $\mathbb{R}^3 \supset \Omega$ - the domain of the terminal bouton; it is assumed that it is a sufficiently regular set;

(ii) $\Omega \supset \Omega_3$ - the domain of production of neurotransmitter;

(iii) $\partial\Omega \supset \partial\Omega_d$ - the regions on the cell membrane in which neurotransmitter is released to the synaptic cleft;

(iv) $\beta : \Omega \to \mathbb{R}$ models the efficiency of the source of neurotransmitter; in the simplest case it can be defined as constant in the domain of neurotransmitter production and as equal to zero outside this domain, i.e. $\beta(x) = 0$ outside Ω_3 and $\beta(\mathbf{x}) = \beta_z$ on Ω_3;

(v) $\bar{\varrho}$ is the balance concentration of vesicles with neurotransmitter inside the bouton; new vesicles can appear only if the concentration is below the balance concentration;

(vi) α denotes the coefficient of the rate of neurotransmitter exocytosis, i.e. it is the number of vesicles released through the unit area of the membrane in unit time by the unit difference of the concentration in the cell and outside the cell; in some types of biological neurons a single action potential activates about 300 vesicles and a single vesicle contains $10^3 \div 10^4$ molecules of neurotransmitter;

(vii) $a_{ij} : \Omega \to \mathbb{R}$ is the diffusion tensor for the vesicles; in the discussed model it is additionally assumed that, in the context of transport, the interior of the bouton is both time independent and homogenous, as well as isotropic; in a such case the tensor is diagonal, independent on space and constant in time; furthermore, the values of all three entries on the diagonal are the same and, as a consequence, the tensor can be reduced to the single coefficient of diffusion; for instance, for the acetylcholine, it is equal to $300\,\mu\,m^2/s$ - see [57]);

(viii) τ denotes the time period through which the neurotransmitter is released from the release regions to the synaptic cleft; in the case of fast transport it is equal to $2 \div 5\,\mu s$;

(ix) t_0 is the moment in which the action potential arrives.

The function $\varrho : \Omega \times [0, T] \to \mathbb{R}$, which denotes the concentration of the vesicles, is the unknown of the model that is based on the diffusion-type equation:

$$\frac{\partial\varrho(\mathbf{x}, t)}{\partial t} = \sum_{i,j=1}^{3} \frac{\partial}{\partial x_i}\left(a_{ij}(\mathbf{x})\frac{\partial\varrho(\mathbf{x}, t)}{\partial x_j}\right) + \beta(\mathbf{x})(\bar{\varrho} - \varrho(\mathbf{x}, t))^+. \qquad (7.12)$$

Fig. 7.5 Domains of the
PDE problem

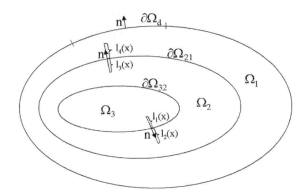

The first term on the right hand side of the equation describes diffusion and the second one models production of neurotransmitter. The function β, which describes production, is weighted by the factor $(\bar{\varrho} - \varrho(x, t))^+$. Therefore, if concentration of the neurotransmitter is greater than the threshold value $\bar{\varrho}$, then he production is stopped. The production term causes the nonlinearity of the Eq. (7.12).

Let us define the boundary conditions (Fig. 7.5).

- Vesicles can leave the bouton only via docking sites

$$\sum_{i,j=1}^{3} a_{ij} \frac{\partial \varrho(\mathbf{x}, t)}{\partial x_j} n_i = 0 \quad \text{for} \quad (\mathbf{x}, t) \in (\partial\Omega - \partial\Omega_d) \times [0, T], \qquad (7.13)$$

- Vesicles can leave the bouton only in the time interval in which the calcium channels are open by the action potential

$$\sum_{i,j=1}^{3} a_{ij} \frac{\partial \varrho(\mathbf{x}, t)}{\partial x_j} n_i = 0 \quad \text{for} \quad (\mathbf{x}, t) \in \partial\Omega_d \times ([0, t_0) \cup (t_0 + \tau, T]). \qquad (7.14)$$

- The vesicles release outside the bouton is proportional to their concentration inside the bouton. Furthermore, it is assumed that vesicle concentration outside Ω is equal to 0.

$$\sum_{i,j=1}^{3} a_{ij} \frac{\partial \varrho(\mathbf{x}, t)}{\partial x_j} n_i = -\alpha \varrho(\mathbf{x}, t) \quad \text{for} \quad (\mathbf{x}, t) \in \partial\Omega_d \times [t_0, t_0 + \tau] \qquad (7.15)$$

Let us also specify the initial condition

$$\varrho(\mathbf{x}, 0) = \varrho_0(\mathbf{x}) \quad \text{on} \quad \Omega. \qquad (7.16)$$

The solution of the (7.12)–(7.16) can be found only by using numerical approximation. The assumption that the bouton is a three-dimensional sphere is the simplest possibility. The numerical solutions based on more realistic geometric assumptions are also studied. The case in which the bouton geometry for computer simulations is based on the image from optical microscope can be put as an example.

Averaging

Let us derive ODE model of fast synaptic transmission by averaging the PDE model. In the below passage, in which the averaging is described, the three-dimensional Lebesgue measure is denoted by m, whereas two-dimensional boundary Lebesgue measure in $\mathbb{R}^3$ is denoted by m_2. The model obtained by averaging will be referred to A-G model.

Let us assume that all the coefficients are smooth i.e. they are of the class C^∞. Such regularity can be easily obtained by a mollifier function. Namely, if the coefficients a_{ij}, β, α are not smooth, but they are only L^∞ functions, then we replace them with smoothed functions. Thus, let p be a standard mollifier kernel. The coefficient a_{ij} can be replaced by $a_{ij}^s(x) = \int_\Omega p_k(z)a_{ij}(x - z)\, dz$. The other coefficient can be smoothed in the same way. The solution ϱ of the equation with smoothed coefficients is also smooth on $\Omega \times (0, T)$ - see, for instance, [66], Sect. 7.1.3. Additionally, the smoothed functions a_{ij} and β are equal to the original ones apart from small neighborhoods of the boundaries $\partial\Omega_{32}$ and $\partial\Omega_{21}$.

Let $\varepsilon > 0$ be fixed and small. Let $\Omega_{3\varepsilon}$ be such a neighbourhood of the boundary $\partial\Omega_{32}$ that $\Omega_{3\varepsilon} \subset \Omega$ and $m(\Omega_{3\varepsilon}) \leq \varepsilon m_2(\partial\Omega_{32})$. Analogously, let $\Omega_{2\varepsilon}$ be such the neighborhood of the boundary $\partial\Omega_{21}$ that is included in $\Omega_2 \subset \Omega$ and $m(\Omega_{2\varepsilon}) \leq \varepsilon m_3(\partial\Omega_{21})$. Furthermore, $a(x) \equiv a_3$ in $\Omega_3 \setminus \Omega_{3\varepsilon}$, $a(x) \equiv a_2$ in $\Omega_2 \setminus (\Omega_{3\varepsilon} \cup \Omega_{2\varepsilon})$ and $a(x) \equiv a_1$ in $\Omega_1 \setminus \Omega_{2\varepsilon}$. Finally $\beta(x) \equiv \beta$ in $\Omega_3 \setminus \Omega_{3\varepsilon}$ and $\beta(x) \equiv 0$ outside Ω_3.

In the integrals considered below dV denotes an element of volume and $d\Gamma$ denotes an element of surface. Let the averaged variables be defined in the following way:

$$\varrho_3(t) = \frac{\int_{\Omega_3 \setminus \Omega_{3\varepsilon}} \varrho(\mathbf{x}, t)\, dV}{m(\Omega_3 \setminus \Omega_{3\varepsilon})}, \tag{7.17}$$

$$\varrho_2(t) = \frac{\int_{\Omega_2 \setminus (\Omega_{3\varepsilon} \cup \Omega_{2\varepsilon})} \varrho(\mathbf{x}, t)\, dV}{m(\Omega_2 \setminus (\Omega_{3\varepsilon} \cup \Omega_{2\varepsilon}))}, \tag{7.18}$$

$$\varrho_1(t) = \frac{\int_{\Omega_1 \setminus \Omega_{2\varepsilon}} \varrho(\mathbf{x}, t)\, dV}{m(\Omega_1 \setminus \Omega_{2\varepsilon})}. \tag{7.19}$$

Let us apply averaging procedure to the inner domain. The below equation is obtained by integrating (7.12) over Ω_3

$$\int_{\Omega_3} \frac{\partial\varrho(\mathbf{x}, t)}{\partial t}\, dV = \int_{\Omega_3} \sum_{i=1}^{3} \frac{\partial}{\partial x_i}\left(a(\mathbf{x}) \sum_{j=1}^{3} \frac{\partial\varrho(\mathbf{x}, t)}{\partial x_j}\right) dV + \int_{\Omega_3} \beta(\mathbf{x})(\bar{\varrho} - \varrho(\mathbf{x}, t))^+\, dV. \tag{7.20}$$

By using Green formula - see, for instance, [66], Appendix C.2, we obtain

$$\int_{\Omega_3 \setminus \Omega_{3\varepsilon}} \frac{\partial \varrho(\mathbf{x}, t)}{\partial t} \, dV + \int_{\Omega_{3\varepsilon} \cap \Omega_3} \frac{\partial \varrho(\mathbf{x}, t)}{\partial t} \, dV = \int_{\partial \Omega_{32}} a(\mathbf{x}) \frac{\partial \varrho(\mathbf{x}, t)}{\partial \nu} \, d\Gamma +$$

$$+ \int_{\Omega_3 \setminus \Omega_{3\varepsilon}} \beta(\mathbf{x})(\bar{\varrho} - \varrho(\mathbf{x}, t))^+ \, dV + \int_{\Omega_{3\varepsilon} \cap \Omega_3} \beta(\mathbf{x})(\bar{\varrho} - \varrho(\mathbf{x}, t))^+ \, dV, \quad (7.21)$$

where the normal derivative is directed outside Ω_3. Let us introduce the following denotation

$$C := \max \left\{ \sup_{(\mathbf{x}, t) \in \Omega \times [0, T]} \varrho(\mathbf{x}, t), \quad \sup_{(\mathbf{x}, t) \in \Omega \times [0, T]} \frac{\partial \varrho(\mathbf{x}, t)}{\partial t} \right\}.$$

Then

$$\left| \int_{\Omega_3 \setminus \Omega_{3\varepsilon}} \frac{\partial \varrho(\mathbf{x}, t)}{\partial t} \, dV - \int_{\partial \Omega_{32}} a(\mathbf{x}) \frac{\partial \varrho(\mathbf{x}, t)}{\partial \nu} \, d\Gamma - \beta \int_{\Omega_3 \setminus \Omega_{3\varepsilon}} (\bar{\varrho} - \varrho(\mathbf{x}, t))^+ \, dV \right| \le$$
$$\le \varepsilon m_2(\partial \Omega_{32})(C + \beta \bar{\rho} + C\beta)$$

Let us denote
$$D := m_2(\partial \Omega_{32})(C + \beta \bar{\rho} + C\beta).$$

New vesicles are produced only if their concentration is less than the threshold value $\bar{\varrho}$. Therefore, it can be assumed that $\varrho(x, t) \le \bar{\varrho}$ and, as a consequence, the positive operator (superscript $(\cdot)^+$) can be omitted. The term in the boundary integral can be approximated as $\frac{a_2 + a_3}{2} \frac{\varrho_2(t) - \varrho_3(t)}{\varepsilon}$. Then

$$\left| m(\Omega_3 \setminus \Omega_{3\varepsilon}) \frac{d}{dt} \varrho_3(t) - m_2(\partial \Omega_{32}) \frac{a_2 + a_3}{2} \frac{\varrho_2(t) - \varrho_3(t)}{\varepsilon} - \beta m(\Omega_3 \setminus \Omega_{3\varepsilon})(\bar{\varrho} - \varrho_3(t)) \right| \le \varepsilon D,$$
$$(7.22)$$

which simply implies

$$\left| \frac{d}{dt} \varrho_3(t) - \frac{m_2(\partial \Omega_{32})(a_2 + a_3)}{2\varepsilon m(\Omega_3 \setminus \Omega_{3\varepsilon})} (\varrho_2(t) - \varrho_3(t)) - \beta(\bar{\varrho} - \varrho_3(t)) \right| \le \varepsilon \frac{D}{m(\Omega_3 \setminus \Omega_{3\varepsilon})}.$$
$$(7.23)$$

Thus, it has been shown that $\varrho_3(t)$ satisfies the following problem

$$\varrho_3(0) = \frac{\int_{\Omega_3 \setminus \Omega_{3\varepsilon}} \varrho(\mathbf{x}, 0) \, dV}{m(\Omega_3 \setminus \Omega_{3\varepsilon})}, \quad (7.24)$$

$$\frac{d\varrho_3(t)}{dt} = \frac{m_2(\partial \Omega_{32})(a_2 + a_3)}{2\varepsilon m(\Omega_3 \setminus \Omega_{3\varepsilon})} (\varrho_2(t) - \varrho_3(t)) + \beta(\bar{\varrho} - \varrho_3(t)) + f_3(t), \quad (7.25)$$

where $|f_3(t)| \le \varepsilon \frac{D}{m(\Omega_3 \setminus \Omega_{3\varepsilon})}$.

After substitutions $E = \bar{\varrho}$, $\frac{1}{R_3 C_3} = \beta$ and $\frac{1}{R_2 C_3} = \frac{m_2(\partial \Omega_{32})(a_2 + a_3)}{2\varepsilon m(\Omega_3 \setminus \Omega_{3\varepsilon})}$ it can be easy noticed that the function ϱ_3 satisfies the Eq. (7.10) with the additional control term f_3.

If $\varepsilon > 0$ is close to zero then the absolute value of the term f_3 becomes close to zero as well. The term $\frac{m_2(\partial \Omega_{32})(a_2 + a_3)}{2\varepsilon m(\Omega_3 \setminus \Omega_{3\varepsilon})}$ in Eq. (7.25), however, becomes large. Therefore, the passing to the limit $\varepsilon \to 0$ causes the singularity in the first term of the right side in Eq. (7.25).

Let us apply averaging procedure to the middle domain. The below equation is obtained by integrating Eq. (7.12) over Ω_2

$$\int_{\Omega_2} \frac{\partial \varrho(\mathbf{x}, t)}{\partial t} \, dV = \int_{\Omega_2} \sum_{i=1}^{3} \frac{\partial}{\partial x_i} \left(a(\mathbf{x}) \sum_{j=1}^{3} \frac{\partial \varrho(\mathbf{x}, t)}{\partial x_j} \right) dV. \tag{7.26}$$

The following is obtained in the analogous way as in the case of the inner domain

$$\int_{\Omega_2 \setminus (\Omega_{3\varepsilon} \cup \Omega_{2\varepsilon})} \frac{\partial \varrho(\mathbf{x}, t)}{\partial t} \, dV + \int_{\Omega_2 \cap \Omega_{3\varepsilon}} \frac{\partial \varrho(\mathbf{x}, t)}{\partial t} \, dV + \int_{\Omega_2 \cap \Omega_{2\varepsilon}} \frac{\partial \varrho(\mathbf{x}, t)}{\partial t} \, dV =$$

$$= \int_{\partial \Omega_{32}} a(\mathbf{x}) \frac{\partial \varrho(\mathbf{x})}{\partial \nu} \, d\Gamma + \int_{\partial \Omega_{21}} a(\mathbf{x}) \frac{\partial \varrho(\mathbf{x})}{\partial \nu} \, d\Gamma, \tag{7.27}$$

where the normal derivative is directed outside the set Ω_2. The first term on the right side of Eq. (7.27) can be approximated as $\frac{a_2 + a_3}{2} \frac{\varrho_3(t) - \varrho_2(t)}{\varepsilon}$, whereas the second one as $\frac{a_1 + a_2}{2} \frac{\varrho_1(t) - \varrho_2(t)}{\varepsilon}$. Thus

$$|m(\Omega_2 \setminus (\Omega_{3\varepsilon} \cup \Omega_{2\varepsilon})) \frac{d}{dt} \varrho_2(t) - m_2(\partial \Omega_{32}) \frac{a_2 + a_3}{2} \frac{\varrho_3(t) - \varrho_2(t)}{\varepsilon}$$

$$- m_2(\partial \Omega_{21}) \frac{a_1 + a_2}{2} \frac{\varrho_1(t) - \varrho_2(t)}{\varepsilon}| \le$$

$$\le \varepsilon C (m_2(\partial \Omega_{32}) + m_2(\partial \Omega_{21})). \tag{7.28}$$

After division by $m(\Omega_2 \setminus (\Omega_{3\varepsilon} \cup \Omega_{2\varepsilon}))$ it is obtained

$$|\frac{d}{dt} \varrho_2(t) - \frac{m_2(\partial \Omega_{32})(a_2 + a_3)}{2\varepsilon m(\Omega_2 \setminus (\Omega_{3\varepsilon} \cup \Omega_{2\varepsilon}))} (\varrho_3(t) - \varrho_2(t))$$

$$- \frac{m_2(\partial \Omega_{21})(a_1 + a_2)}{2\varepsilon m(\Omega_2 \setminus (\Omega_{3\varepsilon} \cup \Omega_{2\varepsilon}))} (\varrho_1(t) - \varrho_2(t))| \le$$

$$\le \varepsilon \frac{C(m_2(\partial \Omega_{32}) + m_2(\partial \Omega_{21}))}{m(\Omega_2 \setminus (\Omega_{3\varepsilon} \cup \Omega_{2\varepsilon}))}. \tag{7.29}$$

Thus, the function $\varrho_2(t)$ satisfies the following problem

$$\varrho_2(0) = \frac{\int_{\Omega_2 \setminus (\Omega_{3\varepsilon} \cup \Omega_{2\varepsilon})} \varrho(\mathbf{x}, 0) \, dV}{m(\Omega_2 \setminus (\Omega_{3\varepsilon} \cup \Omega_{2\varepsilon}))},$$

$$\frac{d}{dt} \varrho_2(t) = \frac{m_2(\partial \Omega_{32})(a_2 + a_3)}{2\varepsilon m(\Omega_2 \setminus (\Omega_{3\varepsilon} \cup \Omega_{2\varepsilon}))}(\varrho_3(t) - \varrho_2(t)) +$$

$$\frac{m_2(\partial \Omega_{21})(a_1 + a_2)}{2\varepsilon m(\Omega_2 \setminus (\Omega_{3\varepsilon} \cup \Omega_{2\varepsilon}))}(\varrho_1(t) - \varrho_2(t)) + f_2(t), \tag{7.30}$$

where $|f_2(t)| \leq \varepsilon \frac{C(m_2(\partial \Omega_{32}) + m_2(\partial \Omega_{21}))}{m(\Omega_2 \setminus (\Omega_{3\varepsilon} \cup \Omega_{2\varepsilon}))}$ and C has been introduced above.

Analogously to the inner pool, in comparison to A-G model (see also Fig. 7.7), where the capacity of the "middle" pool is given by the Eq. (7.9), we can set $\frac{1}{R_1 C_2} = \frac{m_2(\partial \Omega_{21})(a_1 + a_2)}{2\varepsilon m(\Omega_2 \setminus (\Omega_{3\varepsilon} \cup \Omega_{2\varepsilon}))}$ and $\frac{1}{R_2 C_2} = \frac{m_2(\partial \Omega_{32})(a_2 + a_3)}{2\varepsilon m(\Omega_2 \setminus (\Omega_{3\varepsilon} \cup \Omega_{2\varepsilon}))}$. Thus, the function ϱ_2 satisfies the Eq. (7.9) with the additional control term f_2.

Let us apply averaging procedure to the release domain. The below equation is obtained by integrating the Eq. (7.12) over Ω_1

$$\int_{\Omega_1} \frac{\partial \varrho(\mathbf{x}, t)}{\partial t} \, dV = \int_{\Omega_1} \sum_{i=1}^{3} \frac{\partial}{\partial x_i} \left(a(\mathbf{x}) \sum_{j=1}^{3} \frac{\partial \varrho(\mathbf{x}, t)}{\partial x_j} \right) \, dV. \tag{7.31}$$

By using Green formula

$$\int_{\Omega_1 \setminus \Omega_{2\varepsilon}} \frac{\partial \varrho(\mathbf{x}, t)}{\partial t} \, dV + \int_{\Omega_1 \cap \Omega_{2\varepsilon}} \frac{\partial \varrho(\mathbf{x}, t)}{\partial t} \, dV =$$

$$= \int_{\partial \Omega_{21}} a(\mathbf{x}) \frac{\partial \varrho(\mathbf{x}, t)}{\partial \nu} \, d\Gamma + \int_{\partial \Omega_d} a(x) \frac{\partial \varrho(x, t)}{\partial \nu} \, d\Gamma. \tag{7.32}$$

The first term on the right side of (7.32) can be approximated by $\frac{a_1 + a_2}{2} \frac{\varrho_2(t) - \varrho_1(t)}{\varepsilon}$ (the sign "minus" appears since the outer normal is directed inside the domain Ω_1. By utilizing the boundary condition it is obtained

$$m(\Omega_1 \setminus \Omega_{2\varepsilon}) \frac{d\varrho_1(t)}{dt} + \int_{\Omega_1 \cap \Omega_{2\varepsilon}} \frac{\partial \varrho(\mathbf{x}, t)}{\partial t} \, dV =$$

$$\frac{a_1 + a_2}{2} \frac{\varrho_2(t) - \varrho_1(t)}{\varepsilon} m_2(\partial \Omega_{21}) - s(t)\alpha \int_{\partial \Omega_d} \varrho(\mathbf{x}, t) \, d\Gamma. \tag{7.33}$$

The function $s(t) = 1$ during the release period and $s(t) = 0$ otherwise. It can be assumed that the concentration in the release domain does not change significantly in space and is equal to $\varrho_1(t)$. Then

$$\left| m(\Omega_1 \setminus \Omega_{2\varepsilon}) \frac{d\varrho_1(t)}{dt} - \frac{a_1 + a_2}{2} \frac{\varrho_2(t) - \varrho_1(t)}{\varepsilon} m_2(\partial \Omega_{21}) + s(t)\alpha \varrho_1(t) m_2(\partial \Omega_d) \right|$$

$$\leq \varepsilon C m_2(\partial \Omega_{21}), \tag{7.34}$$

where C is defined above.

The formula can be divided by $m(\Omega_1 \setminus \Omega_{2\varepsilon})$

$$
\left| \frac{d\varrho_1(t)}{dt} - \frac{m_2(\partial\Omega_{21})(a_1 + a_2)}{2\varepsilon m(\Omega_1 \setminus \Omega_{2\varepsilon})}(\varrho_2(t) - \varrho_1(t)) + s(t)\frac{\alpha m_2(\partial\Omega_d)}{m(\Omega_1 \setminus \Omega_{2\varepsilon})}\varrho_1(t) \right|
$$
$$
\leq \varepsilon \frac{Cm_2(\partial\Omega_{21})}{m(\Omega_1 \setminus \Omega_{2\varepsilon})}. \tag{7.35}
$$

Thus, the function $\varrho_1(t)$ satisfies the problem

$$
\varrho_1(0) = \frac{\int_{\Omega_1\setminus\Omega_{2\varepsilon}} \varrho(\mathbf{x}, 0)\, dV}{m(\Omega_1 \setminus \Omega_{2\varepsilon})},
$$
$$
\frac{d}{dt}\varrho_1(t) = \frac{m_2(\partial\Omega_{21})(a_1 + a_2)}{2\varepsilon m(\Omega_1 \setminus \Omega_{2\varepsilon})}(\varrho_2(t) - \varrho_1(t))
$$
$$
-s(t)\frac{\alpha m_2(\partial\Omega_d)}{m(\Omega_1 \setminus \Omega_{2\varepsilon})}\varrho_1(t) + f_1(t), \tag{7.36}
$$

where $|f_1(t)| \leq \varepsilon \frac{Cm_2(\partial\Omega_{21})}{m(\Omega_1\setminus\Omega_{2\varepsilon})}$.

Similarly to the cases of the middle and inner pools it can be set $\frac{1}{R_1 C_1} = \frac{m_2(\partial\Omega_{21})(a_1+a_2)}{2\varepsilon m(\Omega_1\setminus\Omega_{2\varepsilon})}$ and $\frac{1}{R_0 C_1} = \frac{\alpha m_2(\partial\Omega_d)}{m(\Omega_1\setminus\Omega_{2\varepsilon})}$. Thus, the function ϱ_1 satisfies the Eq. (7.8) with the additional control term f_1.

Let us notice that the speed of neurotransmitter release is given as $y(t) = s(t)\frac{\alpha m_2(\partial\Omega_d)}{m(\Omega_1\setminus\Omega_{2\varepsilon})}\varrho_1(t)$. Thus, the total amount of the released neurotransmitter during the time interval (t_1, t_2) can be calculated as $N(t_1, t_2) = \int_{t_1}^{t_2} y(t)dt$.

Let us consider controllability, observability and stability of the obtained ODE model and A-G model. By averaging, two time-invariant systems with control have been received. The first one describes the dynamics without release, i.e. the case $s(t) = 0$. It is defined by Eqs. (7.37) and (7.38):

$$
\frac{d\mathbf{z}(t)}{dt} = \mathbf{A}_1\mathbf{z}(t) + \mathbf{B}\mathbf{v}(t), \tag{7.37}
$$

and

$$
\mathbf{y}(t) = \mathbf{C}_1\mathbf{z}(t). \tag{7.38}
$$

The second one describes the dynamics with release, i.e. the case $s(t) = 1$, and it is defined by Eqs. (7.39) and (7.40):

$$
\frac{d\mathbf{z}(t)}{dt} = \mathbf{A}_2\mathbf{z}(t) + \mathbf{B}\mathbf{v}(t), \tag{7.39}
$$

$$
\mathbf{y}(t) = \mathbf{C}_2\mathbf{z}(t). \tag{7.40}
$$

In the above equations

$$\mathbf{z}(t) = [\varrho_1(t),\, \varrho_2(t),\, \varrho_3(t)]^T \text{ and } \mathbf{v}(t) = [f_1(t),\, f_2(t),\, f_3(t) + \beta\bar{\varrho}]^T, \qquad (7.41)$$

the matrix $\mathbf{A}_1$ has the following form

$$\begin{pmatrix} -\frac{m_2(\partial\Omega_{21})(a_1+a_2)}{2\varepsilon m(\Omega_1\backslash\Omega_{2\varepsilon})} - \frac{\alpha m_2(\partial\Omega_d)}{m(\Omega_1\backslash\Omega_{2\varepsilon})} & \frac{m_2(\partial\Omega_{21})(a_1+a_2)}{2\varepsilon m(\Omega_1\backslash\Omega_{2\varepsilon})} & 0 \\ \frac{m_2(\partial\Omega_{21})(a_1+a_2)}{2\varepsilon m(\Omega_2\backslash(\Omega_{3\varepsilon}\cup\Omega_{2\varepsilon}))} & -\frac{m_2(\partial\Omega_{32})(a_2+a_3)}{2\varepsilon m(\Omega_2\backslash(\Omega_{3\varepsilon}\cup\Omega_{2\varepsilon}))} - \frac{m_2(\partial\Omega_{21})(a_1+a_2)}{2\varepsilon m(\Omega_2\backslash(\Omega_{3\varepsilon}\cup\Omega_{2\varepsilon}))} & \frac{m_2(\partial\Omega_{32})(a_2+a_3)}{2\varepsilon m(\Omega_2\backslash(\Omega_{3\varepsilon}\cup\Omega_{2\varepsilon}))} \\ 0 & \frac{m_2(\partial\Omega_{32})(a_2+a_3)}{2\varepsilon m(\Omega_3\backslash\Omega_{3\varepsilon})} & -\frac{m_2(\partial\Omega_{32})(a_2+a_3)}{2\varepsilon m(\Omega_3\backslash\Omega_{3\varepsilon})} - \beta \end{pmatrix},$$

$$(7.42)$$

and the matrix $\mathbf{A}_2$ is of the form

$$\begin{pmatrix} -\frac{m_2(\partial\Omega_{21})(a_1+a_2)}{2\varepsilon m(\Omega_1\backslash\Omega_{2\varepsilon})} & \frac{m_2(\partial\Omega_{21})(a_1+a_2)}{2\varepsilon m(\Omega_1\backslash\Omega_{2\varepsilon})} & 0 \\ \frac{m_2(\partial\Omega_{21})(a_1+a_2)}{2\varepsilon m(\Omega_2\backslash(\Omega_{3\varepsilon}\cup\Omega_{2\varepsilon}))} & -\frac{m_2(\partial\Omega_{32})(a_2+a_3)}{2\varepsilon m(\Omega_2\backslash(\Omega_{3\varepsilon}\cup\Omega_{2\varepsilon}))} - \frac{m_2(\partial\Omega_{21})(a_1+a_2)}{2\varepsilon m(\Omega_2\backslash(\Omega_{3\varepsilon}\cup\Omega_{2\varepsilon}))} & \frac{m_2(\partial\Omega_{32})(a_2+a_3)}{2\varepsilon m(\Omega_2\backslash(\Omega_{3\varepsilon}\cup\Omega_{2\varepsilon}))} \\ 0 & \frac{m_2(\partial\Omega_{32})(a_2+a_3)}{2\varepsilon m(\Omega_3\backslash\Omega_{3\varepsilon})} & -\frac{m_2(\partial\Omega_{32})(a_2+a_3)}{2\varepsilon m(\Omega_3\backslash\Omega_{3\varepsilon})} - \beta \end{pmatrix}.$$

$$(7.43)$$

The other matrices and vectors in formulae (7.37)–(7.40) are given as follows:

$$\mathbf{B} = I_{3\times 3}, \qquad (7.44)$$

$$\mathbf{C}_1 = \begin{bmatrix} 0,\, 0,\, 0 \end{bmatrix}, \quad \mathbf{C}_2 = \left[\frac{\alpha m_2(\partial\Omega_d)}{m(\Omega_1\backslash\Omega_{2\varepsilon})},\, 0,\, 0 \right]. \qquad (7.45)$$

Since the functions f_1, f_2, f_3 are both lower and upper bounded, the set $\mathbb{R}^3 \supset V = V_1 \times V_2 \times V_3$ of accessible controls has the components from the following closed intervals

$$V_1 = \left[-\varepsilon\frac{Cm_2(\partial\Omega_{21})}{m(\Omega_1\backslash\Omega_{2\varepsilon})},\, \varepsilon\frac{Cm_2(\partial\Omega_{21})}{m(\Omega_1\backslash\Omega_{2\varepsilon})} \right],$$

$$V_2 = \left[-\varepsilon\frac{C(m_2(\partial\Omega_{32})+m_2(\partial\Omega_{21}))}{m(\Omega_2\backslash(\Omega_{3\varepsilon}\cup\Omega_{2\varepsilon}))},\, \varepsilon\frac{C(m_2(\partial\Omega_{32})+m_2(\partial\Omega_{21}))}{m(\Omega_2\backslash(\Omega_{3\varepsilon}\cup\Omega_{2\varepsilon}))} \right], \qquad (7.46)$$

$$V_3 = \left[-\varepsilon\frac{D}{m(\Omega_3\backslash\Omega_{3\varepsilon})} + \beta\bar{\varrho},\, \varepsilon\frac{D}{m(\Omega_3\backslash\Omega_{3\varepsilon})} + \beta\bar{\varrho} \right].$$

The model postulated by Aristizabal and Glavinovic - Eqs. (7.8)–(7.11) - define two time-invariant systems with control. The dynamics described by Eqs. (7.47) and (7.48)

$$\frac{d\mathbf{U}(t)}{dt} = \mathbf{A}_3\mathbf{U}(t) + \mathbf{B}\mathbf{v}(t), \qquad (7.47)$$

$$\mathbf{y}(t) = \mathbf{C}_3\mathbf{z}(t), \qquad (7.48)$$

takes place when there is no release, i.e. the case $s(t) = 0$. The system generated by Eqs. (7.49) and (7.50)

$$\frac{d}{dt}\mathbf{U}(t) = \mathbf{A}_4\mathbf{U}(t) + \mathbf{B}\mathbf{v}(t),\tag{7.49}$$

$$\mathbf{y}(t) = \mathbf{C}_4\mathbf{U}(t),\tag{7.50}$$

describes the release time, i.e. the case $s(t) = 1$.

In Eqs. (7.8)–(7.11)

$$\mathbf{v}(t) = \left[0, 0, \frac{E}{R_3C_3}\right]^T \text{ and } \mathbf{U}(t) = [U_1(t), U_2(t), U_3(t)]^T,$$

where U_i is the voltage on the capacitance C_i - see Fig. 7.4, whereas the matrices are given as

$$\mathbf{A}_3 = \begin{pmatrix} -\left(\frac{1}{R_1C_1} + \frac{1}{R_0C_1}\right) & \frac{1}{R_1C_1} & 0 \\ \frac{1}{R_1C_2} & -\left(\frac{1}{R_1C_2} + \frac{1}{R_2C_2}\right) & \frac{1}{R_2C_2} \\ 0 & \frac{1}{R_2C_3} & -\left(\frac{1}{R_2C_3} + \frac{1}{R_3C_3}\right) \end{pmatrix},\tag{7.51}$$

$$\mathbf{A}_4 = \begin{pmatrix} -\frac{1}{R_1C_1} & \frac{1}{R_1C_1} & 0 \\ \frac{1}{R_1C_2} & -\left(\frac{1}{R_1C_2} + \frac{1}{R_2C_2}\right) & \frac{1}{R_2C_2} \\ 0 & \frac{1}{R_2C_3} & -\left(\frac{1}{R_2C_3} + \frac{1}{R_3C_3}\right) \end{pmatrix},\tag{7.52}$$

$$\mathbf{B} = I_{3\times3},\tag{7.53}$$

$$\mathbf{C}_3 = \begin{bmatrix} 0 & 0 & 0 \end{bmatrix}, \quad \mathbf{C}_4 = \begin{bmatrix} \frac{1}{R_0C_1} & 0 & 0 \end{bmatrix}.\tag{7.54}$$

To sum up, the matrices $\mathbf{A}_j$, $j \in \{1, 2, 3, 4\}$ have the form

$$\mathbf{A} = \begin{pmatrix} -(\delta_2 + s(t)\delta_1) & \delta_2 & 0 \\ \delta_3 & -(\delta_3 + \delta_4) & \delta_4 \\ 0 & \delta_5 & -(\delta_5 + \delta_6) \end{pmatrix},\tag{7.55}$$

where $s(t) \in \{0, 1\}$ and $\delta_i > 0$, $i \in \{1, 2, 3, 4, 5, 6\}$.

Theorem 7.3.1 *The dynamical systems with control (7.37), (7.39), (7.47) and (7.49) have the following controllability properties:*

1. *they are uniformly controllable, controllable, controllable at t_0 and controllable in $[t_0, t_1]$,*
2. *they are locally V-controllable to zero provided that $\beta\bar{\varrho} \leq \varepsilon\frac{D}{m(\Omega_3 \setminus \Omega_{3c})}$,*
3. *they are CH(V)-controllable to zero provided that $\beta\bar{\varrho} \leq \varepsilon\frac{D}{m(\Omega_3 \setminus \Omega_{3c})}$,*
4. *they are not V-controllable to zero,*

5. *in the norm* $\|(a_{ij})_{i,j\in\{1,2,3\}}\| := \sum_{i,j=1}^{3} |a_{ij}|$ *the controllability margin of the considered systems is greater or equal to* 1.

Proof In all Eqs. (7.37), (7.39), (7.47) and (7.49) the matrix $\mathbf{B} = I_{3\times3}$. Therefore,

$$\text{rank}\left[\mathbf{B}|\mathbf{AB}|\ldots|\mathbf{A}^{n-1}\mathbf{B}\right] = 3.$$

The controllability of the considered systems is implied by Theorem 5.8.3, whereas their uniform controllability, controllability in $[t_0, t_1]$ and controllability at t_0 are implied by Lemma 5.8.2. If $\beta\varrho \le \varepsilon\frac{D}{m(\Omega_3\backslash\Omega_{3e})}$, then $0 \in V_3$ which implies $0 \in V$. CH(V)-controllability to zero and local V-controllability to zero are obtained by Theorems 5.8.6 and 5.8.5, respectively.

The necessary condition of V-controllability to zero is not satisfied because if $\delta_2 = \delta_3$ and $\delta_4 = \delta_5$, then $\mathbf{A}$ has only real eigenvalues as a symmetric real matrix.

In order to prove the last point of the theorem let us notice that we need to demonstrate that the minimal value of $\|I_{3\times3} - M\|$ is equal to 1 if $\mathbf{M}$ is a singular matrix. Thus, the controllability margin will be grater or equal to 1 because the term that is generated by the matrices $\mathbf{A}_i$ cannot be negative - see Theorem 5.8.9. It is obvious that the matrix

$$\mathbf{K} = \begin{pmatrix} 0 & 0 & 0 \\ 0 & 1 & 0 \\ 0 & 0 & 1 \end{pmatrix} \tag{7.56}$$

is singular and $\|I_{3\times3} - \mathbf{K}\| = 1$. Let us assume, by contradiction, that for some singular matrix $\mathbf{M}$ there is $\|I_{3\times3} - \mathbf{M}\| < 1$.

The matrix $\mathbf{M}$ can be represented as

$$\mathbf{M} = \begin{pmatrix} 1 - \alpha_1 & \beta_1 & \gamma_1 \\ \beta_2 & 1 - \alpha_2 & \gamma_2 \\ \beta_3 & \gamma_3 & 1 - \alpha_3 \end{pmatrix}. \tag{7.57}$$

Then, we have $\|\mathbf{M} - I_{3\times3}\| = \sum_{i=1}^{3}(|\alpha_i| + |\beta_i| + |\gamma_i|)$. By Gershgorin Theorem, for the singular matrix $\mathbf{M}$ its zero eigenvalue must lie within a closed ball $B(1 - \alpha_{i_0}, |\beta_{i_0}| + |\gamma_{i_0}|)$ for some $i_0 \in \{1, 2, 3\}$. If $1 - \alpha_{i_0} < 0$, then $\alpha_{i_0} > 1$ and, as a consequence, $\|\mathbf{M} - I_{3\times3}\| > 1$, which is a contradiction. If $1 - \alpha_{i_0} \ge 0$, then $1 - \alpha_{i_0} - |\beta_{i_0}| - |\gamma_{i_0}| \le 0$ and, as a consequence, $1 \le \alpha_{i_0} + |\beta_{i_0}| + |\gamma_{i_0}| \le \|\mathbf{M} - I_{3\times3}\|$ which is a contradiction as well. $\square$

Theorem 7.3.2 *The systems* (7.39)–(7.40) *and* (7.49)–(7.50) *are observable, The systems* (7.37)–(7.38) *and* (7.47)–(7.48) *are not observable.*

Proof The matrix $\left[\mathbf{C}_i^T|\mathbf{A}_i^T\mathbf{C}_i^T|\left(\mathbf{A}_i^2\right)^T\mathbf{C}_i^T\right]$, $i \in \{2, 4\}$, is a 3×3 upper triangle matrix without zeroes on the diagonal which is implied by (7.55) and the fact that $\mathbf{C}_i^T = [c_1, 0, 0]$, where $c_1 > 0$, $i \in \{2, 4\}$. Therefore,

$$\text{rank}\left[\mathbf{C}_i^T|\mathbf{A}_i^T\mathbf{C}_i^T|\left(\mathbf{A}_i^2\right)^T\mathbf{C}_i^T\right] = 3.$$

The observability of the systems (7.39)–(7.40) and (7.49)–(7.50) is implied by Theorem 5.8.8 whereas lack of the observability of the systems (7.37)–(7.38) and (7.47)–(7.48) is implied by the same theorem due to the fact that $\mathbf{C}_i^T = [0, 0, 0]$, where $i \in \{2, 4\}$. $\qquad\square$

Theorem 7.3.3 *The systems described by* (7.37), (7.39), (7.47) *and* (7.49) *are asymptotically stable.*

Proof Let us calculate Λ_1, Λ_2 and Λ_3 - see Theorem 5.3.5. By using (7.55):

$$\Lambda_1 = -(s(t)\delta_1 + \delta_2 + \delta_3 + \delta_4 + \delta_5 + \delta_6),$$
$$\Lambda_2 = s(t)\delta_1\delta_3 + s(t)\delta_1\delta_4 + \delta_2\delta_4 + \delta_3\delta_5 + \delta_3\delta_6 + \delta_4\delta_6 + s(t)\delta_1\delta_5 + s(t)\delta_1\delta_6 + \delta_2\delta_5 + \delta_2\delta_6,$$
$$\Lambda_3 = -(s(t)\delta_1\delta_3\delta_5 + s(t)\delta_1\delta_3\delta_6 + s(t)\delta_1\delta_4\delta_6 + \delta_2\delta_4\delta_6).$$

Since all the terms of Λ_3 exist also in the term $\Lambda_1\Lambda_2$, this term can be written in the following form: $-\Lambda_1\Lambda_2 = -\Lambda_3 + \kappa$, where $\kappa > 0$, which means that $-\Lambda_1\Lambda_2 + \Lambda_3 = \kappa$. To sum up, the following is satisfied

$$-\Lambda_1 > 0,$$
$$-\Lambda_1\Lambda_2 + \Lambda_3 > 0,$$
$$-\Lambda_3 > 0.$$

The thesis is obtained by Theorems 5.3.5 and 5.3.1. $\qquad\square$

7.3.3 Model of Neuropeptide Slow Transport

In the model of slow neuropeptide transport, similarly as in the model of fast transport, the transport process is described by using diffusion-type equations to represent both the space dependencies such as gradients of concentration, localization of release regions and ion channels and changes in time. The model describes the LDCVs activation, diffusion, accumulation and release. This model, based on partial differential equations, can be the starting point for the ordinary differential model that can be obtained from PDE by averaging. The relations between these two models are also discussed in the sequel. Furthermore, on the basis of the ordinary differential model, the electronic circuit model is proposed and discussed.

Let us specify notations and assumptions that concern the domains corresponding to functionally specific regions of the bouton - see also Fig. 7.6:

- $\mathbb{R}^n \supset \Omega$, $n \in \{2, 3\}$ is the closed set that represents the synaptic bouton; for simulations in 2D and 3D n is equal to 2 and 3 respectively.
- Ω_1 is the closed set that is the part of the domain Ω in which inactive LDCVs are accumulated. The set $\Omega \setminus \Omega_1$ denotes the central region of the bouton that is devoid of inactive vesicles.

Fig. 7.6 An example of a configuration of the sets used in the PDE model. Ω_1 is the set in which immobile vesicles accumulate while $\Omega \setminus \Omega_1$ is devoid of them. The boundary section $\partial\Omega_{Ca}$ represents the location of calcium channels while $\partial\Omega_N$ is the part of the boundary through which new vesicles arrive

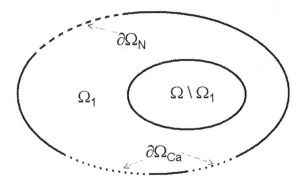

- $\partial\Omega_1$ denotes the boundary of the set $\Omega \setminus \Omega_1$ that is contained in the interior of the set Ω. The inactive LDCVs are accumulated in the neighborhood of the boundary $\partial\Omega$.
- $\partial\Omega_{Ca}$ is the part of the boundary of the set Ω in which the calcium channels are situated, $\partial\Omega_{Ca} \subsetneq \partial\Omega$.
- $\partial\Omega_N$ denotes the part of the boundary of the set Ω through which inactive LDCVs enter the bouton, $\partial\Omega_N \subsetneq \partial\Omega$. The set $\partial\Omega_N$ models the connection of bouton with the body of the neuron. It is assumed that $\partial\Omega_N \cap \partial\Omega_1 = \emptyset$ which means that the calcium channels cannot be located at the entrance boundary.
- ν denotes the outer normal versor to the boundary of the considered sets. It is assumed that the sets are sufficiently smooth i.e. that ν exists in each of the points of boundary.

Let us specify the model assumptions.

- The variables $\rho_N(\mathbf{x}, t)$, $\rho_A(\mathbf{x}, t)$ and $\rho_{Ca}(\mathbf{x}, t)$, that denotes the concentrations of inactive LDCVs, active LDCVs and ions of calcium respectively, are the unknowns of the model.
- Both the active vesicles and the ions diffuse freely in the domain Ω. The diffusion coefficients are denoted as μ_{Ca} and μ_A, respectively.
- The inactive vesicles can move only in the domain Ω_1. Moreover, the intensity of the diffusion varies between the points and depends on the directions i.e. it is anisotropic and heterogeneous. That means, among others, that the diffusion properties cannot be described by a single number but they have to be characterized by the diffusion tensor, denoted as $\mu_N^{ij}(\mathbf{x})$, $i, j \in \{1, \ldots, n\}$.
- Vesicles are activated if the concentration of calcium ions locally exceed the threshold value ρ_{Ca}^{Thr}. If this condition is satisfied, then the dependence between the intensity of activation and the ions concentration is linear. The constants δ, β and α denote the reaction rates for the inactive vesicles, the active vesicles and the calcium ions, respectively.
- The inactive vesicles arrive to the bouton, which is modelled by Ω, via $\partial\Omega_N$. The speed of the arrival is proportional to the difference between E_N and the current local concentration of the inactive vesicles, with the coefficient of the proportion

denoted by η. The inactive vesicles can be accumulated if their local concentration is less than E_N.

- The calcium ions enter the bouton if the two following conditions are satisfied simultaneously: If their value of concentration is less than the balance value E_{Ca} and if the ion channels are open. The action potential opens the channels that are modelled by the boundary set $\partial\Omega_{Ca}$. The channel capacity is described by the function $f(t)$ which is equal to the rate of the flow if the channel is open. Otherwise, the value of the function is equal to 0.
- The active vesicles cannot dock at the bouton boundary which is the connection with the neuron body. The release rate of the neuropeptide depends linearly with the coefficient γ on the concentration of the vesicles.
- The initial distributions ρ_{N0}, ρ_{A0}, ρ_{Ca0} of the concentrations ρ_N, ρ_A and ρ_{Ca}, respectively, are given.

The reaction-diffusion problem, modeled by PDEs, is given by the following equations.

(PN) The dynamics of the inactive neuropeptide is described as the following problem:

$$\frac{\partial\rho_N(\mathbf{x}, t)}{\partial t} = \frac{\partial}{\partial x_j}\mu_N^{ij}(\mathbf{x})\frac{\partial\rho_N(\mathbf{x}, t)}{\partial x_i} -$$

$$-\delta(\rho_{Ca}(\mathbf{x}, t) - \rho_{Ca}^{Thr})^+\rho_N(\mathbf{x}, t) \text{ on } \Omega_1, \tag{7.58}$$

$$\rho_N(\mathbf{x}, t) = 0 \text{ on } \Omega \setminus \Omega_1, \tag{7.59}$$

$$\mu_N^{ij}(x)\frac{\partial\rho_N(\mathbf{x}, t)}{\partial x_i}\nu_j = \eta(E_N - \rho_N(\mathbf{x}, t))^+ \text{ on } \partial\Omega_N, \tag{7.60}$$

$$\mu_N^{ij}(\mathbf{x})\frac{\partial\rho_N(\mathbf{x}, t)}{\partial x_i}\nu_j = 0 \text{ on } \partial\Omega_1 \setminus \partial\Omega_N, \tag{7.61}$$

$$\rho_N(\mathbf{x}, 0) = \rho_{N0}(x) \text{ on } \Omega_1. \tag{7.62}$$

(PA) The dynamics of the active neuropeptide is described as the following problem:

$$\frac{\partial\rho_A(\mathbf{x}, t)}{\partial t} = \mu_A\Delta\rho_A(\mathbf{x}, t) +$$

$$+ \beta(\rho_{Ca}(\mathbf{x}, t) - \rho_{Ca}^{Thr})^+\rho_N(\mathbf{x}, t) \text{ on } \Omega, \tag{7.63}$$

$$\frac{\partial\rho_A(\mathbf{x}, t)}{\partial\nu} = -\gamma\rho_A(\mathbf{x}, t) \text{ on } \partial\Omega \setminus \partial\Omega_N, \tag{7.64}$$

$$\frac{\partial\rho_A(\mathbf{x}, t)}{\partial\nu} = 0 \text{ on } \partial\Omega_N, \tag{7.65}$$

$$\rho_A(\mathbf{x}, 0) = \rho_{A0}(x) \text{ on } \Omega. \tag{7.66}$$

(PCa) The calcium ions dynamics is described as the following problem:

$$\frac{\partial \rho_{Ca}(\mathbf{x}, t)}{\partial t} = \mu_{Ca} \Delta \rho_{Ca}(\mathbf{x}, t) -$$

$$- \alpha(\rho_{Ca}(\mathbf{x}, t) - \rho_{Ca}^{Thr})^+ \rho_N(\mathbf{x}, t) \text{ on } \Omega, \tag{7.67}$$

$$\frac{\partial \rho_{Ca}(\mathbf{x}, t)}{\partial \nu} = f(t)(E_{Ca} - \rho_{Ca}(\mathbf{x}, t))^+ \text{ on } \partial\Omega_{Ca}, \tag{7.68}$$

$$\frac{\partial \rho_{Ca}(\mathbf{x}, t)}{\partial \nu} = 0 \text{ on } \partial\Omega \setminus \partial\Omega_{Ca}, \tag{7.69}$$

$$\rho_{Ca}(\mathbf{x}, 0) = \rho_{Ca0}(\mathbf{x}) \text{ on } \Omega. \tag{7.70}$$

As it has been aforementioned, the ODE model can be obtained from the PDE by averaging its unknowns. Let us introduce the following denotations for the averaged values of ρ_N, ρ_A, ρ_{Ca} over Ω_1 and $\Omega \setminus \Omega_1$.

$$\overline{\rho_N}^1(t) = \frac{\int_{\Omega_1} \rho_N(\mathbf{x}, t) \, dV}{m(\Omega_1)}, \tag{7.71}$$

$$\overline{\rho_A}^1(t) = \frac{\int_{\Omega_1} \rho_A(\mathbf{x}, t) \, dV}{m(\Omega_1)}, \tag{7.72}$$

$$\overline{\rho_A}^2(t) = \frac{\int_{\Omega \setminus \Omega_1} \rho_A(\mathbf{x}, t) \, dV}{m(\Omega \setminus \Omega_1)}, \tag{7.73}$$

$$\overline{\rho_{Ca}}^1(t) = \frac{\int_{\Omega_1} \rho_{Ca}(\mathbf{x}, t) \, dV}{m(\Omega_1)}, \tag{7.74}$$

$$\overline{\rho_{Ca}}^2(t) = \frac{\int_{\Omega \setminus \Omega_1} \rho_{Ca}(\mathbf{x}, t) \, dV}{m(\Omega \setminus \Omega_1)}. \tag{7.75}$$

The defined variables correspond to ρ_N, ρ_{A_R}, ρ_{A_I}, ρ_{Ca_R} and ρ_{Ca_I}, respectively, in the compartment model presented below.

Let us propose the compartment model of the slow neuropeptide transmission. The model is based on a system of ODEs, analogous to the one introduced in [5].

In the model of the slow neuropeptide transport, according to the fact that activation of calcium ions has to be put into consideration explicitly, at least five pools are necessary: the pool of inactive LDCVs, two pools of active LCDVs and two pools for calcium. Both the ions and the active LDCVs can occupy the whole domain of the bouton, whereas the inactive LDCVs can accumulate only near the membrane in the selected regions. Let us introduce the following notations:

$\rho_N(t)$ denotes the concentration of inactive LDCVs in the accumulation region;

$\rho_{A_R}(t)$ denotes the concentration of activated LDCVs in the accumulation region of inactive LDCVs;

$\rho_{A_I}(t)$ denotes the concentration of activated LDCVs in the central region of the bouton devoid of LDCVs;

$\rho_{Ca_R}(t)$ - the concentration of calcium ions in the accumulation region of inactive LDCVs;

$\rho_{Ca_I}(t)$ - the concentration of calcium ions in the central region which is devoid of LDCVs.

The first equation models dynamics of vesicles concentration ρ_N. The dynamics is caused both by activation by calcium ions and by transport from the neuron soma. This transport takes place only if the concentration inside the domain is lower then E_N. The activation causes the decrease of the inactive vesicles and the decrease is proportional to two factors. The difference between the concentration of the calcium threshold value, below which the activation does not occur, and its current concentration is the first factor. The concentration of the inactive vesicles is the second one. The equation has the following form:

$$\frac{d\rho_N(t)}{dt} = p_1(E_N - \rho_N(t))^+ - d_1(\rho_{CaR}(t) - \rho_{Ca}^{Thr})^+ \rho_N(t). \tag{7.76}$$

In Eq. (7.76) the constants d_1 and p_1 describe the reaction rate constant of the vesicle activation process and the flow rate of vesicles into the bouton, respectively. The assumption that the flow rate p_1 depends only on their concentration has been made. It should be stressed, however, that in the light of the results described in [168] the assumption is far simplistic because it is activity dependent. In order to express this dependence, the rate should depend either on the capacity of the calcium pool ρ_{Ca_R} or explicitly on time.

The second equation models the dynamics of concentration of ρ_{A_R} in the activation region. This concentration decreases during the release of the neuropeptide to the synaptic cleft and it increases during the activation of the vesicles by the ions. As it has been aforementioned, the activated vesicles can diffuse freely in the whole domain. The equation has the following form:

$$\frac{d\rho_{A_R}(t)}{dt} = d_2(\rho_{Ca_R}(t) - \rho_{Ca}^{Thr})^+ \rho_N(t) + a_1(\rho_{A_I}(t) - \rho_{A_R}(t)) - p_2\rho_{A_R}(t), \tag{7.77}$$

where a_1 denotes the diffusion rate of active LDCVs in the considered pool in the bouton cytoplasm and d_2 represents the reaction rate constant of the vesicle activation process. The constant p_2 denotes the release rate of LDCVs.

The third equation models the dynamics of the concentration of the active vesicles ρ_{A_I} in the central region of the bouton. This concentration increases after activation because the diffusion process tends to balance the concentration of the active vesicles. The equation has the following form:

$$\frac{d\rho_{A_I}(t)}{dt} = -a_2(\rho_{A_I}(t) - \rho_{A_R}(t)), \tag{7.78}$$

where a_2 represents the diffusion constant of active LDCVs.

Two last equations describe the changes of the calcium ions concentration. The first one models the dynamics of the ions concentration ρ_{Ca_R} of the inactive LDCVs in the region of accumulation. The right side of the equation consists of three components. The first one describes the inflow of the calcium ions through the channels when the channels are open by the action potential. This inflow is proportional to the difference between the current concentration and the balance concentration E_{Ca}.

The second term represents the diffusion of the ions between the central region and the region of accumulation of the inactive LSCVs. The third component represents the calcium utilization during the activation process. The equation has the following form:

$$\frac{d\rho_{Ca_R}(t)}{dt} = g(t)(E_{Ca} - \rho_{Ca_R}(t))^+ +$$
$$+ b_1(\rho_{Ca_I}(t) - \rho_{Ca_R}(t)) - d_3(\rho_{Ca_R}(t) - \rho_{Ca}^{Thr})^+ \rho_N(t), \qquad (7.79)$$

where the constant b_1 represents the diffusion coefficient of calcium ions in the considered pool and the constant d_3 is the reaction rate of the activation process. The function g is defined in the following way:

$$g(t) = \begin{cases} p_3 > 0, & \text{if the membrane is activated by the action potential,} \\ 0, & \text{otherwise,} \end{cases}$$

where the constant p_3 denotes the rate of the inflow rate of calcium from the cleft to the bouton when the voltage dependent channels are open.

The last equation models the dynamics of concentration ρ_{Ca_I} of the calcium in the central region of the bouton. The associated pool fills, by diffusion, with the ions that are not used to activate LDCVs in the pool ρ_N. Then, it is used as the reservoir for calcium. The equation has the following form:

$$\frac{d\rho_{Ca_I}(t)}{dt} = -b_2(\rho_{Ca_I}(t) - \rho_{Ca_R}(t)), \qquad (7.80)$$

where b_2 represents the diffusion coefficient of calcium in the pool.

The structure of the mutual dependencies of Eqs. (7.76)–(7.80) is presented in Fig. 7.7 whereas possible spatial distribution of the pools is shown in Fig. 7.8.

In order to derive the ordinary differential model from the partial differential model let us make the assumption that the values of ρ_N, ρ_A and ρ_{Ca} in points of Ω_1 and $\Omega \setminus \Omega_1$ can be approximated by the averaged values (7.71)–(7.75). According to this assumption the specified concentrations are not varied too much in the averaging domains and, as a consequence, the forces of diffusion are considered only between

Fig. 7.7 The flows between the pools present in the ODE model

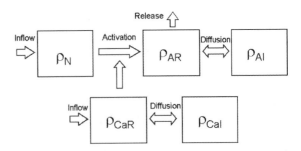

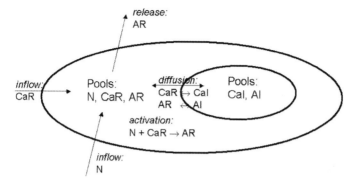

Fig. 7.8 Possible distribution of spatial pools and the placement of processes in the ODE model of slow transport

Table 7.1 Relations between parameters of ODE and PDE models. Inflow/outflow capacities, reaction rate parameters and actual diffusion coefficients are the parameters of PDE model. The parameters of ODE model are the lumped ones which means that they depend on the domain geometry. The parameter h denotes the length scale of the domains Ω_1 and $\Omega \setminus \Omega_1$, m is the Lebesgue measure in $\mathbb{R}^n$ whereas σ denotes the Lebesgue boundary measure

Parameter type	ODE model	PDE model
Threshold concentration	$E_{Ca}, E_N, \rho_{Ca}^{Thr}$	$E_{Ca}, E_N, \rho_{Ca}^{Thr}$
Inflow and Outflow	p_1	$\frac{\eta\sigma(\partial\Omega_N)}{m(\Omega_1)}$
	$g(t)$	$\frac{\mu_{Ca} f(t)\sigma(\partial\Omega_{Ca})}{m(\Omega_1)}$
	p_2	$\frac{\gamma\mu_A\sigma(\partial\Omega\setminus\partial\Omega_N)}{m(\Omega_1)}$
Reaction rate	d_1	δ
	d_2	β
	d_3	α
Diffusion	a_1	$\frac{\mu_A\sigma(\partial\Omega_1\setminus\partial\Omega)}{2hm(\Omega_1)}$
	a_2	$\frac{\mu_A\sigma(\partial\Omega_1\setminus\partial\Omega)}{2hm(\Omega\setminus\Omega_1)}$
	b_1	$\frac{\mu_{Ca}\sigma(\partial\Omega_1\setminus\partial\Omega)}{2hm(\Omega_1)}$
	b_2	$\frac{\mu_{Ca}\sigma(\partial\Omega_1\setminus\partial\Omega)}{2hm(\Omega\setminus\Omega_1)}$

the domains of averaging that correspond to the compartments (pools). The relations between parameters and constants of these two models are summarized briefly in Table 7.1.

Electronic Model

As it has been aforementioned, the modelling of neuronal phenomena by using analog electronic circuits is a standard approach. Two ways of creating such models

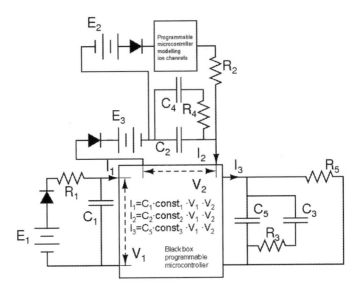

Fig. 7.9 Electric circuit representing the presynaptic episode of slow transmission. The large black box represents the Ca^{2+} activation of LDCVs. The laws that govern the box are placed inside. The smaller box represents the pattern of bouton stimulation

are possible. The first one is possible if ODEs, that describe the modelled process dynamics, are given. Then, the circuit whose dynamics is described by the same ordinary differential equations can be built. The second way is to build the circuit which has similar dynamic properties as the modelled process. In such a case, the system of ODEs can be found. The system describes the dynamics of the circuit and, what follows, the modelled process. In the considered case the first way can be applied. The system of ODEs (7.76)–(7.80) are, however, complex, and therefore it is difficult to reproduce them by a circuit that is made of some elementary electronic modules. Therefore a two-step top-down methodology has been applied. The general structure of the circuit was designed in the first step - see Fig. 7.9. Modelling of the pools is a standard task so the loops that model their dynamics were designed in details. The module of the circuit which models LDCVs activation by calcium ions is represented by a black box. Such an approach was worked out in cybernetics, where black boxes are used to model cybernetic systems that have unknown inner structure. Furthermore, it is assumed that in the black boxes the responses for input signals are known. Thus, Eqs. (7.81)–(7.83) that describe the black-box inner processes dynamics were modelled by using a programmable microcontroller. Another programmable microcontroller allows the system to generate signal pulses of any frequency. In such a way the pulses $\frac{h(t)}{C_2 \cdot R_2}(E_2 - U_{C_2})^+$, used in Eq. (7.86), were modelled.

The electronic circuit that represents the introduced ODEs model is shown in Fig. 7.9. Synthesis of LDCVs, inflow of calcium through the membrane via ion channels and the threshold concentration of calcium are modelled by the sources E_1, E_2, E_3, respectively. The unidirectional flow of current, which is ensured

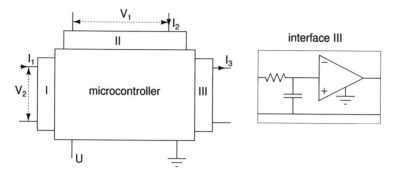

Fig. 7.10 The structure of the black box. A microcontroller is responsible for the multiplication of voltages whereas the interfaces I, II, III enforce the desired output currents. The interface III is shown in details as the example

by diodes, corresponds to the fact that only positive parts of the terms $(E_N - \varrho_{N3})$, $(\varrho_{Ca2} - \bar{\varrho}_{Ca2})$, $(E_{Ca} - \varrho_{Ca1})$ appear in Eqs. (7.84)–(7.86). The voltages U_{C_1}, U_{C_2}, U_{C_3}, U_{C_4}, U_{C_5} on the capacitors represent the mean concentrations ϱ_N, ϱ_{A_I}, ϱ_{Ca_R}, ϱ_{Ca_I} and ϱ_{A_R} of vesicles and ions in Eqs. (7.76)–(7.80). The resistors model flow resistances between the pools. The black box models the reaction between ions and inactive vesicles. This is realized by using an electronic microchip.

Let us assume that the black box is governed by the following formulae (Fig. 7.10)

$$I_1 = C_1 \cdot \alpha_1 \cdot V_1 \cdot V_2, \tag{7.81}$$

$$I_2 = C_2 \cdot \alpha_2 \cdot V_1 \cdot V_2, \tag{7.82}$$

$$I_3 = C_5 \cdot \alpha_3 \cdot V_1 \cdot V_2. \tag{7.83}$$

Then, the following equations are obtained by using Kirchhoff laws

$$\frac{dU_{C_1}}{dt} = \frac{1}{C_1 R_1}(E_1 - U_{C_1})^+ - \alpha_1(U_{C_2} - E_3)^+ U_{C_1}, \tag{7.84}$$

$$\frac{dU_{C_5}}{dt} = \alpha_3(U_{C_2} - E_3)^+ \cdot U_{C_1} + \frac{1}{C_5 R_3}(U_{C_3} - U_{C_5}), \tag{7.85}$$

$$\frac{dU_{C_2}}{dt} = \frac{1}{C_2 \cdot R_4}(U_{C_4} - U_{C_2}) - \alpha_2(U_{C_2} - E_3)^+ + \frac{h(t)}{C_2 R_2}(E_2 - U_{C_2})^+, \tag{7.86}$$

$$\frac{dU_{C_3}}{dt} = -\frac{1}{C_3 R_3}(U_{C_3} - U_{C_5}), \tag{7.87}$$

$$\frac{dU_{C_4}}{dt} = -\frac{1}{R_4 C_4}(U_{C_4} - U_{C_2}), \tag{7.88}$$

where α_i are constant.

To sum up, the circuit dynamics is described by the system of Eqs. (7.84)–(7.88) which corresponds to model (7.76)–(7.80). The dynamics of the black box is modelled by the Eqs. (7.81)–(7.83) and realized by a programmable microchip that has, among others, analog-digital and digital-analog converters and a microprocessor with a memory.

7.4   Model of the Synapse

In general, two types of synapses can be distinguished - the chemical and electric ones (see Chap. 2 for more details). Let us start discussion from the electric synapse. It is far simpler than the chemical one and, what follows, its can be easily modelled.

Let us discuss a model of the electric synapse (gap junction). The properties of transmission, that determine the dynamics of the voltage, are the starting point for the analysis. Then, the electric circuit, in which the changes of voltage have analogous dynamics, is proposed as an electronic model of the synapse. The specification of ODE differential problem, that describes the voltage dynamics in the circuit, is the last step of the model creation.

Referring to the properties of transmission of the signal via the gap junction, let us recall that the postsynaptic potential is both delayed and attenuated in comparison with the presynaptic signal [21, 22, 79]. The cell membrane acts as the capacitor. The aforementioned delay is caused by the fact that the postsynaptic potential is detectable if the postsynaptic membrane capacitance is charged [22]. Therefore, although the ionic current flows through the synapse without delay, the postsynaptic potential is delayed. Thus, the electric synapse acts as the low-pass filter [22, 79]. The simplest electronic circuit which acts as this type of filter is presented in Fig. 7.11 and can be used as a circuit model of the gap junction.

Derivation the mathematical formula which describes the dynamics of voltage in the circuits can be done by using elementary methods. By Kirchhoff current law

$$i_1(t) = i_2(t) + i_3(t), \tag{7.89}$$

whereas Kirchhoff voltage law and Ohm law leads to

$$V_1(t) = R_1 i_1(t) + V_2(t); \text{ thus } i_1(t) = \frac{V_1(t)}{R_1} - \frac{V_2(t)}{R_1}, \tag{7.90}$$

$$i_2(t) = \frac{V_2(t)}{R_2}, \tag{7.91}$$

$$i_3(t) = c\frac{dV_2(t)}{dt}. \tag{7.92}$$

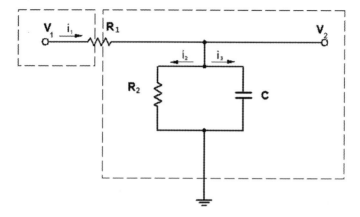

Fig. 7.11 The circuit model of the electric synapse (gap junction). In the left dotted frame there is the fragment which corresponds to presynaptic current. In the right dotted frame there is the fragment which corresponds to the postsynaptic conductance

Putting (7.90)–(7.92) into (7.89) we obtain

$$\frac{V_1(t)}{R_1} - \frac{V_2(t)}{R_1} = \frac{V_2(t)}{R_2} + c\frac{dV_2(t)}{dt} \tag{7.93}$$

and

$$\frac{dV_2(t)}{dt} + \frac{R_1 + R_2}{cR_1R_2}V_2(t) = \frac{1}{cR_1}V_1(t). \tag{7.94}$$

Thus, the dynamics of changes of the voltage in the circuit, which is the model of the gap junction, is described by a linear differential problem of the form

$$V_2(0) = V_0$$
$$\frac{dV_2(t)}{dt} + bV_2(t) = aV_1(t), \tag{7.95}$$

where $b = \frac{R_1+R_2}{cR_1R_2} > 0$ and $a = \frac{1}{cR_1} > 0$.

If an input signal $V_1(t)$ is given as a standard type of spike, it can be approximated as $V_1(t) = a \cdot t \cdot e^{-t}$. Then, the postsynaptic voltage can be easily calculated from (7.95) as $V_2(t) = \frac{a}{b-1}\left(((b-1)t - 1)e^{-t} + e^{-bt}\right)$ - see Fig. 7.12 for $a = 1$ and $b = 2$.

The chemical synapse consists of three parts - the presynaptic bouton, the synaptic cleft and the membrane of the postsynaptic neuron. In general, two ways of the modelling of the chemical synapse can be applied. In the first approach, that leads to an extremely simplified model, the synapse is modelled as one module which means that its inner structure is neglected. Such models exist and they are based on probabilistic approach. They are not discussed in this monograph because probabilistic models are beyond the scope of this monograph. The second approach consists in

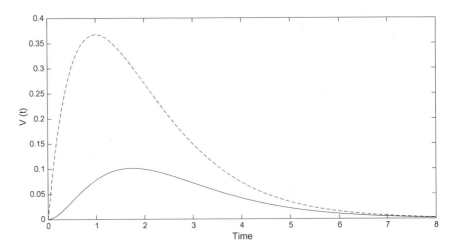

Fig. 7.12 The presynaptic impulse (dotted line), given as $V_1(t) = t \cdot e^{-t}$ and the postsynaptic voltage $V_2(t) = (t - 1)e^{-t} + e^{-2t}$, which is a solution of the problem (7.95)

modelling transport processes in all three parts of the chemical synapse. Such an approach leads, however, to the models of high computational complexity. Therefore, the processes are modelled in each part separately. The models of fast and slow transport in the presynaptic bouton has been presented and discussed in Sect. 7.3. Transport in synaptic cleft has diffusive character and it can be described by a diffusion equation

$$\frac{\partial \varrho(x, y, z, t)}{\partial t} = a \left(\frac{\partial^2}{\partial x^2} + \frac{\partial^2}{\partial y^2} + \frac{\partial^2}{\partial z^2} \right) \varrho(x, y, z, t) + f(x, y, z, t). \quad (7.96)$$

In the above equation ϱ denotes the density of transmitter, a is the coefficient of diffusion and f models the source. The Eq. (7.96) is completed by the initial condition

$$\varrho(x, y, z, 0) = 0 \quad (7.97)$$

and by the Neumann boundary conditions

$$\left. \frac{\partial \varrho}{\partial z} \right|_{z=0} = 0 \quad \text{and} \quad \left. \frac{\partial \varrho}{\partial z} \right|_{z=d} = 0, \quad (7.98)$$

where d denotes the width of the synaptic cleft. According to the above boundary conditions, the transmitter is reflected at the presynaptic and postsynaptic side. The source function f describes the release of the neurotransmitter from one vesicles and can be assumed as $f(x, y, z, t) = \alpha \exp\left(-\left(\frac{x^2+y^2}{b} + \frac{z^2}{c}\right)\right)$, where α, a and b are positive constants. The model describes correctly the diffusion of glutamate in the synaptic cleft.

7.5 Bibliographic Remarks

The models of the membrane fragments can be found in [106, 141], Sect. 2.6 whereas a cable model of the current in the neuron is described in [106, 157], Chap. 8. The application of the linear cable equation for modelling dendritic conductance is discussed in [103, 115, 178].

The use of analog electronic circuits as the models of biological phenomena are a standard approach in biological modelling, including subcellular processes - see [106].

The biological foundations of the pool models are presented in [1].

The A-G model, presented in this monograph was proposed in [5].

The PDE model of neurotransmitter fast transport in the presynaptic neuron is proposed and analysed in [41], numerical simulations in two-dimensional version, based on this model, are presented in [44], whereas some control aspects of the model are discussed in [42]. The PDE model of neuropeptide slow transport as well as the model based on the corresponding electronic circuit can be found in [43]. The formal problems connected with averaging in the context of the model of fast transport are considered in [50].

The electronic model of the electric synapse (gap junction) is described in [22].

Some studies that concern the possibilities of modelling wide spectrum of signal processing in the nervous system, such as interneural long-distance and short-distance signalling and dendro-dendritic coupling, are presented in [49]. In that paper not only basic electronic elements constituted the circuit, as it is usually in the standard studies, but also digital-analog and analog-digital modules were used as components of the modelling circuit.

Cybernetic foundations, including black boxes methodology, can be found in [7], Chap. 6.

Theoretical foundations of mollification procedure is described in [123], Lemma 2.21, p. 50. An example of its application can be found, for instance, in [47], in which it was used in the model of dynamics of the training process of the perceptron (see also Chap. 11 in this book).

The aforementioned probabilistic model of the chemical synapse was proposed by del Castillo and Katz [62] and it was also discussed in [106], Sect. 7.1.1. The model was applied to mammalian neuromuscular junction [51].

Hodgkin–Huxley model was proposed in [94] and it was discussed in [165], Chap. 4 and Appendices F and G as well as in [106], Sect. 4.1 and [175], Sect. 4.5.

The model, which is contemporary called FitzHugh–Naguno model, was introduced by FitzHugh [76, 77]. It was clarified by using a tunnel diode which was introduced in [141]. Various types of the nonlinear module in the FitzHugh–Naguno model were applied in [133]. The circuit modelling FitzHugh equations, in which operational amplifiers were used, was proposed in [105]. The model as such and its dynamical aspects were discussed in detail in [106], Sect. 4.2.

The model of the chemical synapse which is based on kinetic formalism and Markov processes is proposed in [63]. The used formalism enables to describe acting of the voltage-dependent channels in the presynaptic membrane, the release of neurotransmitter to the synaptic cleft and the gating the postsynaptic receptors. As the authors of the article declare: *"This framework can facilitate the integration of a wide range of experimental data and promote consistent theoretical analysis of neural mechanisms from molecular interactions to network computations."*

The model of diffusion in synaptic cleft is discussed in [110]. The model is based on diffusion partial differential equation and was applied effectively to describe the diffusion of glutamine in the synaptic cleft.

Part IV
Mathematical Models of the Perceptron

Chapter 8
General Model of the Perceptron

The general model of the perceptron is presented in this chapter. The model consists of two parts. The first one is a mathematical description of structure of the artificial neural network. The description is based on graph theory and it is very general. It is valid for each type of neural networks, not only for the perceptron. The formal basis of training process of the perceptron is presented in Sect. 8.2. Next, in Sect. 8.3, the training process of the perceptron is discussed in the context of dynamical systems theory.

8.1 Model of a Structure of a Neural Network

As it has been aforementioned, the general approach to a mathematical description of artificial neural networks structure is proposed in this section. Since it is based on graph theory, let us recall some very basic definitions that concern oriented graphs - so called *orgraphs*. In this type of graphs the edges are oriented.

Definition 8.1 Let a finite set A be given. An orgraph G is an ordered pair $G :=$ (A, Ed), where $Ed \subset A \times A$. The set A is the set of the nodes of the graph G, whereas the set Ed is a set of its edges.

Let us set that $(a_i, a_j), \in Ed$ is the edge from the node a_i to the node a_j. Let us notice that, according to the above definition, at most one edge (a_i, a_j) belongs to the set Ed. The oriented graphs in which this condition is not satisfied are called multigraphs. They are not considered in this monograph.

If graphs are used to description of artificial neural network structures, then the nodes denote neurons whereas the edges define connections between them.

Let $\#A$ denotes the power (the number of elements) of the finite set A.

© Springer International Publishing AG, part of Springer Nature 2019
A. Bielecki, *Models of Neurons and Perceptrons: Selected Problems and Challenges*, Studies in Computational Intelligence 770,
https://doi.org/10.1007/978-3-319-90140-4_8

Definition 8.2 The number $\delta_G := \#A$ is called the degree of the graph G.

Definition 8.3 Let a graph $G = (A, Ed)$ be given. The number $\delta_{a_i}^+ := \#\{a_j :$ $(a_j, a_i) \in Ed\}$ is called the input semidegree of the node a_i, whereas the number $\delta_{a_i}^- := \#\{a_j : (a_i, a_j) \in \mathcal{E}\}$ is called the output semidegree of the node a_i.

The above definition means that the input semidegree of the node a_i is the number of the edges that enter the node a_i. Similarly, the output semidegree of the node a_i is the number of the edges that exit from the node a_i.

The untrained deterministic neuron can be defined formally as a function of two vector variables.

Definition 8.4 An untrained neuron which has k inputs and processes the input signals that belong to the set $X \subset \mathbb{R}^k$ (a k neuron over X for abbreviation) is a function of two real variables defined as follows

$$F : \mathbb{R}^k \times X \ni (\mathbf{w}, \mathbf{x}) \to F(\mathbf{w}, \mathbf{x}) = f(<\mathbf{w}, \mathbf{x}>) \in \mathbb{R},$$

where "$< \cdot, \cdot >$" denotes a real scalar product and $f : \mathbb{R} \to \mathbb{R}$ is called an activation function of a neuron. If f is a linear function, then the neuron is called a linear neuron. A learned k-neuron (called also a trained neuron or programmed neuron) over X is a function

$$F^* := F(\mathbf{w}, \cdot) : X^k \to \mathbb{R},$$

where $\mathbf{w} \in \mathbb{R}^k$.

Remarks   1. In practice a standard real scalar product is used, i.e.

$$< \mathbf{w}, \mathbf{x} >= \sum_{k'=1}^{k} w_{k'} \cdot x_{k'} := \mathbf{w} \circ \mathbf{x}.$$

2. Without the loss of generality the identity mapping can be used as an activation function in the linear neuron - see Lemma 6.1.
3. In nonlinear neurons bounded functions are usually used as activation functions.
   Let us assume that the following objects are given.

$G := (A, Ed)$ - an orgraph which has the degree δ_G and such that $\{a \in A : \delta_a^+ = 0\} \neq \emptyset$;

$\gamma : A \ni a \to \gamma(a) \in L$ - a bijective mapping;

$\mathcal{F}$ - the set of all neurons or, alternatively, the set of all trained neurons;

$\alpha : A \ni a \to \alpha(a) \in F$, where $\alpha(a)$ is a k-neuron if $\delta_a^+ = 0$ and $\alpha(a)$ is δ_a^+-neuron otherwise;

$W := \{(l, m) : l \in L, m \in \{1, ..., k\}$ if $\delta_a^+ = 0$ and $m \in \{1, ..., \delta_a^+\}$ if $\delta_a^+ \neq 0$, where $l = \gamma(a)$;

$\beta : \mathcal{E} \to W$ - a bijective mapping.

Definition 8.5 A quintuple

$$\mathcal{S}_k := (G, \gamma, \alpha : A \to F, W, \beta)$$

is said to be the <u>structure of a k-neural network</u> if F denotes the set of all neurons and it is said to be the <u>structure of a learned k-neural network</u> if F is the set of all trained neurons.

Remarks 1. The condition $\{a \in A : \delta_a^+ = 0\} \neq \emptyset$ ensures that in the considered neural network there exist neurons onto which the inputs of the external stimuli are put. The definition does not guarantee, however, that the output neurons exist in the network. In recurrent networks, for instance, the output neurons do not exist - the terminal pattern of the network neurons excitations is treated as the network response.

2. The γ mapping indexes the nodes of the graph by natural numbers.

3. The α mapping designs neurons to the nodes of the graph G. Let us assume that a $k - ANN$ is considered. Then, a $k-$neuron is assigned to a node whose input semidegree is equal to zero. If $\delta_a^+ \neq 0$, then a δ_a^+-neuron is assigned to this node.

4. W is the set of the indices that index the inputs of neurons. The first element l in a pair is the index of a neuron, usually $l \in \{1, ..., \delta_G\}$. In perceptrons, however, L is usually a set of the bi-indices in which the first element encodes the number of a layer whereas the second one encodes the number of a neuron in a given layer.

5. The input of the neuron, to which a given component of the external stimulus or the signal from other neuron is put onto, is determined by the mapping β. Every weight in the neurons that are not the input ones can be identified with an edge of the graph G which describes the ANN structure because β is a bijective mapping.

In this monograph the perceptrons are considered. Their structure is defined as follows.

Definition 8.6 Let
$$\mathcal{S}_k := (G = (A, Ed), \gamma, \alpha, W, \beta)$$

be a structure of a $k-$neural network. Let us also assume that the set A of the nodes of graph G is a disjoint union of nonempty subsets $A_1, ..., A_R$ such that $a_i \in A_r$ and $a_j \in A_{r+1}$, for each edge $(a_i, a_j) \in Ed$ ($r \in \{1, ..., R-1\}$). Let, furthermore, m nodes belong to A_R. Then, $\mathcal{S}_{k,m}$ is called a structure of $R-$ layer $k-$ neural network and the corresponding neural network is called $R - k - m-$ perceptron, and it is denoted as $\mathcal{PRC}_R^{k,m}$. If, furthermore, $(a_i, a_j) \in Ed$ for each $a_i \in A_r$ and $a_j \in A_{r+1}$, then the perceptron is called complete. The trained perceptron is defined in an analogical way - in such a case $\mathcal{S}_k$ is the structure of a trained neural network.

Remarks   1. In perceptrons the activation functions of neurons in the same layer
   are the same. Activation functions in different layers, however, can be different.
   Perceptrons, usually, have two layers and the logistic function is used in hidden
   layer whereas identity is used in output layer. Such solution is justified by the
   theoretical results that concern the approximation properties of perceptrons. In
   the neural networks that have not the layer structure activation functions of all
   neurons are usually the same.
2. By definition, in perceptrons the neurons that belong to the first layer are the
   input neurons whereas the neurons of the last layer are the output ones.
3. In complete perceptrons an output signal from a neuron of the rth layer is put
   onto each neuron of the $r + 1-$st layer. This implies that the number of inputs
   of the neurons in the $r + 1-$st layer is equal to the number of neurons in the rth
   layer.
4. The nodes that belong to A_R correspond to the output neurons of the perceptron.

   The defined above structure of a perceptron is the basis for formal definition of
the perceptron.

Definition 8.7 Let a structure

$$\mathcal{S}_k := (G, \gamma, \alpha, W, \beta)$$

of a $\mathcal{PRC}_R^{k,m}$ be given, where $G = (A, E)$. Let sets $X \subset \mathbb{R}^k$ and $Y \subset \mathbb{R}^m$ be such
sets that for each input signal $\mathbf{x} \in X$ the corresponding output signal $\mathbf{y}$ of $\mathcal{PRC}_R^k$
belongs to Y. A three-tuple

$$\mathcal{PRC}_R{}^{k,m}(X, Y) := (X, \mathcal{PRC}_R^{k,m}, Y)$$

is called an $\underline{R - k - m \text{ perceptron over } X}$ trained or not depending on the status of
the considered perceptron.

Remarks   1. An example of a complete $PRC_2^{5,2}$ is shown in Fig. 3.1 whereas an
   incomplete perceptron is shown in Fig. 8.1.
2. The set X is the set of input stimuli whereas Y contains all output signals that
   correspond to the stimuli from X.
3. The definition of a recurrent neural network can be put forward in analogous
   way. The only difference concerns the set Y. In the case of a recurrent network
   it is an infinite sequence of vectors. A number of the components of vectors is
   equal to the number of all neurons in the considered network. Thus, a vector
   represents a single excitation of the network and the sequence is a chain of the
   whole network excitations. An example of a recurrent neural network is shown
   in Fig. 8.2.

   Let a neural network $\mathcal{PRC}_R^{k,m}(X, Y)$ be given. When the vector stimulus $\mathbf{x} = (x_1, ..., x_k) \in X$ is put into the input layer of the perceptron, the signals appear on
the outputs of all m neurons of the output layer. Their output signal of the whole

Fig. 8.1 An example of an
uncomplete perceptron

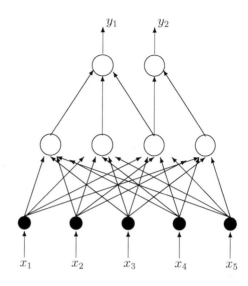

Fig. 8.2 An example of
structure of a recurrent
neural network

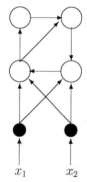

perceptron can be considered as a vector $(y_1, ..., y_m) = \mathbf{y} \in Y \subset \mathbb{R}^m$. Let $\{(\mathbf{x}, \mathbf{y}(\mathbf{x})) :$
$\mathbf{x} \in X,\ \mathbf{y}(\mathbf{x}) \in Y\}$ be a set of all pairs such that $\mathbf{y}(\mathbf{x})$ is the output signal of the
network of the perceptron if the vector $\mathbf{x}$ is put into the network input. In such
a way the function $\mathfrak{F} : X \to Y$ has been constructed. The function corresponds to
$\mathcal{PRC}_R^{k,m}(X, Y)$ and it can be identified with the perceptron.

The following corollary is implied by the above construction.

Corollary 8.8 *Let us assume that a trained perceptron* $\mathcal{PRC}_1^{k,m}(X, Y)$ *consists of*
T neurons, which means that $Y \subset \mathbb{R}^T$, is given. Let the functions

$$F_1, ..., F_T : \ F_t : \mathbb{R}^k \supset X \to Y \subset \mathbb{R},\ t \in \{1, ..., T\}$$

correspond to the trained neurons of the perceptron. Then, the function $\mathfrak{F}$, corre-
sponding to the perceptron, has the following form

$$\mathfrak{F} : X \ni \mathbf{x} \to \mathfrak{F}(\mathbf{x}) = (F_1(\mathbf{x}), ..., F_T(\mathbf{x})) \in Y.$$

Let us assume that two trained perceptrons $\mathcal{PRC}_{R_1}^{k,m}(X, Y)$ and $\mathcal{PRC}_{R_2}^{k,m}(X, Y)$ are given. Let, furthermore, functions $\mathfrak{F}_1 : X \to Y$ and $\mathfrak{F}_2 : X \to Y$ correspond to $\mathcal{PRC}_{R_1}^{k,m}(X, Y)$ and $\mathcal{PRC}_{R_2}^{k,m}(X, Y)$, respectively.

Definition 8.9 If the functions $\mathfrak{F}_1 : X \to Y$ and $\mathfrak{F}_2 : X \to Y$ are equal identically, then the perceptrons are called <u>equivalent</u>.

The above definition implies that if two equivalent perceptrons are treated as black boxes, then they cannot be differentiated because for each stimulus their reactions are the same.

The definition of equivalent perceptrons implies the following corollary.

Corollary 8.10 *Let structures*

$$\mathcal{S}_k := (G = (A, Ed), \gamma, \alpha, W, \beta)$$

and

$$\mathcal{S}_k^\star := (G^\star = (A^\star, Ed^\star), \gamma^\star, \alpha^\star, W^\star, \beta^\star),$$

correspond to the incomplete trained perceptron $\mathcal{PRC}_{R_1}^{k,m}(X, Y)$ and to the complete trained perceptron $\mathcal{PRC}^{\star k,m}_{R_1}(X, Y)$, respectively. Let $A^\star = A$ and $Ed \subset Ed^\star$. If the weights of the the perceptron $\mathcal{PRC}^{\star k,m}_{R_1}(X, Y)$ that correspond to the edges which do not belong to Ed are all equal to zero and all the remaining weights of $\mathcal{S}_k^\star$ are equal to the corresponding weights of $\mathcal{S}_k$, then the perceptrons $\mathcal{PRC}_{R_1}^{k,m}(X, Y)$ and $\mathcal{PRC}^{\star k,m}_{R_1}(X, Y)$ are equivalent.

Let us notice that the above corollary implies that, without the loss of generality, only complete perceptrons can be considered in theoretical studies.

Let us consider two trained perceptrons: $\mathcal{PRC}_{R_1}^{k,m}(X, Y)$ and $\mathcal{PRC}_{R_2}^{m,l}(Y, Z)$ with the corresponding functions $\mathfrak{F}_1 : \mathbb{R}^k \ni X \to Y \in \mathbb{R}^m$ and $\mathfrak{F}_2 : Y \to Z \in \mathbb{R}^l$. Let us also assume that the output signal of $\mathcal{PRC}_{R_1}^{k,m}(X, Y)$ is put onto the input of $\mathcal{PRC}_{R_2}^{m,l}(Y, Z)$. In such a way a new perceptron $\mathcal{PRC}_{R_1+R_2}^{k,l}(X, Z)$ is obtained. This construction can be described formally in the following way.

Definition 8.11 Let the structures $\mathcal{S}_k$ and $\mathcal{S}_k^\star$ of the trained perceptrons $\mathcal{PRC}_{R_1}^{k,m}$ (X, Y) and $\mathcal{PRC}_{R_2}^{m,l}(Y, Z)$ have the form

$$\mathcal{A}_k := (G = (A, \mathcal{E}), \gamma, \alpha, W, \beta)$$

and

$$\mathcal{A}_m^* := (G^* = (A^*, \mathcal{E}^*), \gamma^*, \alpha^*, W^*, \beta^*),$$

respectively. Let us construct a new trained neural network $\mathcal{PRC}_{R_1+R_2}^{k,l}(X, Z)$ which has the structure

$$\tilde{\mathcal{A}}_k := (\tilde{G} = (\tilde{A}, \tilde{\mathcal{E}}), \tilde{\gamma}, \tilde{\alpha}, \tilde{W}, \tilde{\beta})$$

obtained in the following way:

$\tilde{A} := A \cup A^*$,

$\tilde{\mathcal{E}} := \mathcal{E} \cup \mathcal{E}^* \cup \{(a_i, a_j) : a_i \in A, \ \delta_{a_i}^- = 0, \ a_j \in A^*, \ \delta_{a_j}^+ = 0\}$,

$\tilde{\gamma} : \tilde{A} \ni a \to \tilde{\gamma}(a)$ - a bijection,

$\tilde{\alpha} : \tilde{A} \ni a \to \tilde{\alpha}(a) \in F$ is such a mapping that $\tilde{\alpha}|A \equiv \alpha$ and $\tilde{\alpha}|A^* \equiv \alpha^*$,

$\tilde{\beta} : \tilde{\mathcal{E}} \to \tilde{W}$ is a bijection and $\tilde{W}$ is a set of indices created according to Definition 8.5.

The perceptron $\mathcal{PRC}_{R_1+R_2}^{k,l}(X, Z)$ constructed in the way described above is called a <u>superposition</u> of the perceptrons $\mathcal{PRC}_{R_1}^{k,m}(X, Y)$ and $\mathcal{PRC}_{R_2}^{m,l}(Y, Z)$.

The definitions of a perceptron and a superposition of perceptrons imply directly two following corollaries.

Corollary 8.12 *A multilayer perceptron is a superposition of one-layer perceptrons it consists of.*

Corollary 8.13 *Let a perceptron $\mathcal{PRC}_R^{k,l}(X, Z)$ with the corresponding function $\mathfrak{F} : X \to Z$ be superposition of the perceptrons $\mathcal{PRC}_{R_1}^{k,m}(X, Y)$ and $\mathcal{PRC}_{R_2}^{m,l}(Y, Z)$ with corresponding functions $\mathfrak{F}_1 : X \to Y$ and $\mathfrak{F}_2 : Y \to Z$, respectively. Then*

$$\mathfrak{F} = \mathfrak{F}_2 \circ \mathfrak{F}_1.$$

Let us consider the following illustrating example. Let a signal $\mathbf{x} \in \mathbb{R}^k$ be put on the input of the trained perceptron $\mathcal{PRC}_R^{n_0,n_m}(X, Y)$. The signal is processed simultaneously by the neurons of the first layer of the perceptron. Provided that the first layer consists of n_1 neurons, the first layer realizes the function

$$g_1 : \mathbb{R}^{n_0} \ni X \to \mathbb{R}^{n_1}.$$

In the case of the exemplary perceptron shown in Fig.3.1, $\mathbf{x} \in \mathbb{R}^5$ and

$$g_1 : \mathbb{R}^5 \to \mathbb{R}^4.$$

The output signal $\mathbf{s}$ of the first layer is put onto the input of the second layer which processes it and creates the output signal of the second layer. This process is continued until the output signal from the penultimate layer is processed by the last layer that create its output signal which is the output signal of the whole perceptron. In the case of the aforementioned exemplary perceptron

$$g_2 : \mathbb{R}^4 \to \mathbb{R}^2,$$

and the function corresponding to it has the form:

$$\mathfrak{F} = g_2 \circ g_1 : \mathbb{R}^5 \ni X \to Y \in \mathbb{R}^2.$$

In general, a trained perceptron $\mathcal{PRC}_R^{k,m}(X, Y)$ realizes the function

$$\mathfrak{F} = g_R \circ g_{R-1} \circ \cdots \circ g_1 : \mathbb{R}^k \to \mathbb{R}^m.$$

To sum up, a trained perceptron $\mathcal{PRC}_R^{k,m}(X, Y)$ acts in such a way that after putting a stimulus $\mathbf{x} \in X \subset \mathbb{R}^k$ to the input layer it creates the reaction $\mathbf{y} \in Y \subset \mathbb{R}^m$.

8.2   Supervised Deterministic Training Process

In this section a formal approach to supervised deterministic training process of perceptrons is presented.

A trained perceptron should solve a given problem. This means that in response to a given stimulus $\mathbf{x}$ the perceptron reaction $\mathbf{y}$ should be equal to the desired (correct) value, let us say, $\mathbf{z}$. If the values of some of the desired output signals are known, then the training sequence, used in the sequel in the supervised training process, can be created.

Let us define a training set.

Definition 8.14 A finite sequence of the pairs

$$\left(\mathbf{x}^{(1)}, \mathbf{z}^{(1)} \right), ..., \left(\mathbf{x}^{(N)}, \mathbf{z}^{(N)} \right),$$

where $\mathbf{x}^{(n)} \in X \subset \mathbb{R}^k$, $\mathbf{z}^{(n)} \in Y \subset \mathbb{R}^m$, $n \in \{1, ..., N\}$ is called a <u>training set</u> of the perceptron $\mathcal{PRC}_R^{k,m}(X, Y)$ if for each $n \in \{1, ..., N\}$ the vector $\mathbf{z}^{(n)}$ is the correct reaction of the perceptron to the stimulus $\mathbf{x}^{(n)}$.

The training process of the perceptron consists in setting such weights in all neurons that the difference between the perceptron reaction $\mathbf{y}^{(n)}$ and the correct reaction $\mathbf{z}^{(n)}$ is as small as possible. This problem can be formulated as finding a minimum of a certain function.

Let us assume that an untrained perceptron $\mathcal{PRC}_R^{k,m}(X, Y)$ and its training set $(\mathbf{x}^{(1)}, \mathbf{z}^{(1)}), ..., (\mathbf{x}^{(N)}, \mathbf{z}^{(N)})$ are given.

Definition 8.15 A function

$$E : \mathbb{R}^{s_1} \ni \mathbf{w} \to E(\mathbf{w}) \in [0, \infty)$$

is called an <u>error function</u> if it is of the form

$$U(\mathbf{y}^{(1)}(\mathbf{w}), ..., \mathbf{y}^{(N)}(\mathbf{w})),$$

where

$$U : (\mathbb{R}^{M_R})^N \to [0, \infty) \text{ and } \mathbf{y}^{(n)} : \mathbb{R}^{s_1} \to \mathbb{R}^{M_R},$$

and s_1 is a number of all weights in the perceptron whereas M_R is a number of neurons in the output layer. Furthermore, it is required that fulfilling the following system of equations

$$\mathbf{y}^{(1)}(\mathbf{w}) = \mathbf{z}^{(1)}, ..., \mathbf{y}^{(N)}(\mathbf{w}) = \mathbf{z}^{(N)},$$

is the necessary and sufficient condition of zeroing the function E at the point $\mathbf{w}$.

Remarks 1. In the above definition $\mathbf{w}$ denotes the vector of all weights of the considered perceptron. As it has been aforementioned, the perceptron weights and, as a consequence, the components of $\mathbf{w}$ can be indexed in various way. In this mongraph they are used interchangeably depending on the convenience in a given context. Anyway, the components of $\mathbf{w}$ are arranged in a certain way.

2. Vectors $\mathbf{y}^{(n)}(\mathbf{w}) \in Y$ are output signals of the perceptrons that are the reactions to input signals $\mathbf{x}^{(n)}$ which are the elements of the training set. It should be stressed that during the training process the weights in neurons change and, as a consequence, at various stages of the training process the perceptron reactions to the same input signal are various, as well.

3. Iterative algorithms in which derivatives are used are common methods of supervised learning of perceptrons. Therefore, some additional assumptions concerning regularity of the function E are often specified. It is assumed, usually, that E is at least of class C^1.

4. Theoretically, finding the global minimum of the function E would be the best solution of the problem of a perceptron training. Nevertheless, it is impossible to solve it by using analytical methods - it would be a system of equations whose number is equal to the number of all weights in the perceptron. Furthermore, in the case of a nonlinear perceptron, the equations would be nonlinear as well. Therefore, it is looking for a sufficiently good local minimum by using numerical methods. Difference iterative schemata, in which the error function E is used explicitly, are commonly applied. The descent gradient method with application of back-propagation is the simplest one.

5. The mathematical analysis of the aforementioned schemata has not only numerical aspect but also it has the optimization and dynamical aspects. The last one is analyzed in this monograph.

6. The finding weights that zeroes the function E, which corresponds to the finding the global minimum of E, although theoretically optimal, is not optimal in practice because of the so called *overtraining*. In such a case a trained perceptron reacts perfectly to stimuli from the training set but, usually, it does not react correctly to the stimuli that do not belong to the training set.

8.3   Gradient Learning Process

In this section the gradient descent method, which is the simplest algorithm in the group of differential training methods of the perceptron, is considered. Back propagation technique that is necessary in the case of the multilayer perceptron is utilized in the training process.

In order to analyze the descent gradient learning process formally, let us assume that the training set $\{\mathbf{x}^{(n)}, \mathbf{z}^{(n)}\}_{n=1,2,...,N}$ of a perceptron $\mathcal{PRC}_R^{k,m}(X, Y)$ is given. Let, furthermore, the activation function of each neuron be of a C^1 class. Let a training algorithm be given by a descent gradient iterative scheme

$$\mathbf{w}^{(p+1)} := \mathbf{w}^{(p)} - h \cdot \operatorname{grad} E(\mathbf{w}^{(p)}), \tag{8.1}$$

which can be also presented for each component separately

$$w_j^{(p+1)} := w_j^{(p)} - h \cdot \frac{\partial E(\mathbf{w}^{(p)})}{\partial w_j}, \quad j \in \{1, ..., J\}, \tag{8.2}$$

where J is the number of all weights in the perceptron and $h \in (0, 1)$. The variable p corresponds to the number of a step of iteration. It can be easily noticed that the presented schema is the Euler method applied to the following gradient differential equation

$$\frac{d\mathbf{w}}{dt} = -\operatorname{grad} E(\mathbf{w}). \tag{8.3}$$

This observation is the starting point for the analysis presented in this monograph - the training process, described by formulae (8.1) and (8.2), will be studied by using dynamical systems theory referred to the gradient differential equation (8.3).

In the sequel the following function, so called square error function, is used as the error function

$$E(\mathbf{w}) = \sum_{n=1}^{N} \sum_{t=1}^{T_R} \left[y_{R,t}^{(n)} - z_t^{(n)} \right]^2, \tag{8.4}$$

where $y_{R,t}^{(n)} = f_{R,t}\left(\mathbf{w}_{R,t} \circ \mathbf{x}_{R,t}^{(n)}\right)$, $f_{R,t}$ is a C^1 activation function of the tth neuron in the Rth layer, the input signal of the penultimate layer is put into the input of the last layer $\mathbf{x}_{R,t}(n) = \mathbf{y}_{R-1}$, $\mathbf{w}_{R,t} = [w_{R,t,1}, \ldots, w_{R,t,m}]$ and the following convention of indexing is used:

$n = 1, \ldots, N$ - the number of element in the training set specified as superscript in brackets,
$r = 1, \ldots, R$ - the number of a layer, $t = 1, \ldots, T_r$ - the number of a neuron in the rth layer,
$m = 1, \ldots, M_r$ - the number of an input of a neuron in the rth layer.

As it has been already mentioned $T_{r-1} = M_r$.

To sum up, the following convention of indexing of outputs and weights of neurons is used:

$y_{r,t}^{(p)}$ - the output signal of the tth neuron in the rth layer in the pth training step,

$w_{r,t,m}^{(p)}$ - the mth weight in the tth neuron in the rth layer in the pth training step,

$\mathbf{w}_{r,t}$ - the vector of weights of the tth neuron of the rth layer.

Thus, for instance, $y_{2,7}^{(11)}$ denotes the output signal of the seventh neuron in the second layer in the eleventh training step.

The error function E depends, explicitly, only on weights of the neurons that belong to the last layer of the perceptron and on the input signals of the last layer. These signals are the output signals of the penultimate layer and they depend explicitly on the weights of the neurons of this layer and on the input signals of this layer. Thus, the error function E depends on all the weights of the perceptron. Since the input signals $\mathbf{x}^{(n)}$ are fixed as the elements of the training set, the vector of all weights of the perceptron is the only variable on which the function E is dependent.

In order to apply the formula (8.2) efficiently all the gradient components $\frac{\partial E}{\partial w_j}$, where $j \in \{1, \ldots, J\}$ indexes all the weights of a perceptron, should be calculated. Let us calculate the change of weights of the neurons that belong to the last layer. The output signal of the perceptron depends directly on these weights, so

$$y_{R,t}^{(p)} = f_{R,t}\left(\sum_{m=1}^{M_R} x_{R,m}^{(p)} \cdot w_{R,t,m}^{(p)}\right),$$

where $x_{R,m}^{(p)} = y_{R-1,m}^{(p)}$. Thus, by formula (8.2), the weights change rule for the last layer has the form

$$w_{R,t,m}^{(p+1)} = w_{R,t,m}^{(p)} - \eta \cdot \frac{\partial E}{\partial w_{R,t,m}}.$$

Let us assume that the square error function is used. Then

$$\frac{\partial E}{\partial w_{R,t_0,m_0}} = \frac{\partial}{\partial w_{R,t_0,m_0}} \sum_{n=1}^{N}\sum_{t=1}^{T_R}\left[f_{R,t}\left(\sum_{m=1}^{M_R} y_{R-1,m}^{(n)} \cdot w_{R,t,m}^{(n)}\right) - z_t^{(n)}\right]^2 =$$

$$= \sum_{n=1}^{N}\frac{\partial}{\partial w_{R,t_0,m_0}} \sum_{t=1}^{T_R}\left[f_{R,t}\left(\sum_{m=1}^{M_R} y_{R-1,m}^{(n)} \cdot w_{R,t,m}^{(n)}\right) - z_t^{(n)}\right]^2 =$$

$$= 2 \cdot \sum_{n=1}^{N}\left[y_{R,t_0}^{(n)} - z_{t_0}^{(n)}\right] \cdot f'_{R,t_0}\left(s_{R,t_0}^{(n)}\right) \cdot y_{R-1,m_0}^{(n)}$$

where $s_{R,t_0}^{(n)} := \sum_{m=1}^{M_R} y_{R-1,m}^{(n)} \cdot w_{R,t,m}^{(n)}$ is a total excitation of the t_0th neuron of the Rth layer in the nth step and f' denotes the derivative of the function f.

In order to calculate the weights changes for neurons of the hidden layers, the chain rule must be applied because the input signal **y** of the perceptron depends on them indirectly. Thus, for the penultimate layer

$$w_{R-1,t,m}^{(p+1)} = w_{R-1,t,m}^{(p)} - \eta \cdot \frac{\partial E}{\partial w_{R-1,t,m}}$$

$$\frac{\partial E}{\partial w_{R-1,t,m}} = \sum_{n=1}^{N} \sum_{t=1}^{T_{R-1}} \frac{\partial E}{\partial y_{R-1,t}^{(n)}} \cdot \frac{\partial y_{R-1,t}^{(n)}}{\partial w_{R-1,t,m}}.$$

The calculations can be continued in the same way as for the neurons of the output layer.

For the subsequent layers calculations can be done in analogous way. The presented way of calculations of the weights changes is called back-propagation method.

8.4 Bibliographic Remarks

The presented general description of a neural network structure is based on the approach presented in [32]. Other examples of graph theory applications to description of the structure of artificial neural networks can be found in [153, 154]. The description of the structure of perceptrons is based on the formalism presented in [188].

The back-propagation training method of the perceptrons was worked out independently by a few groups of scientists [64, 181] and was described, for instance, in [90], Sect. 6.1, [139], Sect. 6.2.

Chapter 9
Linear Perceptrons

In this chapter the linear perceptrons are considered. In Sect. 9.1 some basic properties of structures of the linear perceptrons are discussed whereas in Sect. 9.2 the dynamics of the training process of the linear perceptrons is analyzed. The stability of the training process is studied in Sect. 9.3.

9.1 Basic Properties of Linear Perceptrons

Let us start from the basic facts that have crucial significance in the theory of linear perceptrons.

Corollary 9.1.1 *Each programmed linear neuron is equivalent to a linear neuron that has the identity function as its activation function.*

Proof Because of clarity, in order to specify all subtleties, in the proof the scalar product is denoted as $< \cdot, \cdot >$.

Let us assume that the considered linear $M-$neuron has a corresponding function $F : \mathbb{R}^M \times \mathbb{R}^M \to \mathbb{R}$. Then, the function corresponding to the trained neuron, which has the weight $\mathbf{w}$, is of the form $F = f(< \cdot, \mathbf{w} >)$, where $f : \mathbb{R} \to \mathbb{R}$ is the activation function which is linear i.e. $f(x) = a \cdot s$, $a \in \mathbb{R}$ and $s :=< \mathbf{x}, \mathbf{w} >$ is the total excitation of the neuron. The trained neuron which has the corresponding function $F^\star(\mathbf{x}) =< \mathbf{x}, a \cdot \mathbf{w} >$ is equivalent to the considered one because

$$F(\mathbf{x}) = f(< \mathbf{x}, \mathbf{w} >) = a < \mathbf{x}, \mathbf{w} >=< \mathbf{x}, a \cdot \mathbf{w} >= F^\star(\mathbf{x}) = \mathrm{id}(< \cdot, a \cdot \mathbf{w} >).$$

$\square$

© Springer International Publishing AG, part of Springer Nature 2019
A. Bielecki, *Models of Neurons and Perceptrons: Selected Problems and Challenges*, Studies in Computational Intelligence 770,
https://doi.org/10.1007/978-3-319-90140-4_9

Thus, without loss of generality, it can be assumed that linear neurons have identity as activation function.

The following corollary is simply implied by the properties of scalar product.

Corollary 9.1.2 *A linear function* $F : \mathbb{R}^m \supset X \to \mathbb{R}$ *corresponds to a linear* M*−neuron with the set* X *as its set of input signals.*

It turns out that, without loss of generality, only one-layer linear perceptrons can be considered.

Lemma 9.1.3 *Each multilayer linear perceptron is equivalent to a one-layer linear perceptron.*

Proof Since a multilayer perceptron consists of a finite number of layers, it is sufficient to prove the lemma only for two layers. By Corollary 9.1.1, it can be assumed that all neurons have identity as activation function.

Let us consider a neuron which belongs to an output layer. Let this layer consists of M_2 neurons.

$$y_{2,t} = \mathbf{y_1} \circ \mathbf{w_{2,t}},$$

where $\mathbf{y}_1$ is a vector which is formed by an output signals of the neurons from the first layer and $\mathbf{w}_{2,t}$ is a weight vector of the tth neuron of the second layer. The vector $\mathbf{y}_1$ can be presented in an orthogonal basis $\{\mathbf{e}_1, \ldots, \mathbf{e}_{M_2}\}$

$$\mathbf{y}_1 = \sum_{m=1}^{M_2} y_{1_m} \mathbf{e}_m = \sum_{m=1}^{M_2} (\mathbf{x} \circ \mathbf{w}_{1,m}) \mathbf{e}_m,$$

where $\mathbf{x}$ is an input signal of the perceptron. By the properties of scalar product we have

$$y_{2,t} = \sum_{m=1}^{M_2} ((\mathbf{x} \circ \mathbf{w}_{1,m}) \mathbf{e}_m) \circ \mathbf{w}_{2,t} = \sum_{m=1}^{M_2} (\mathbf{x} \circ \mathbf{w}_{1,m})(\mathbf{e}_m \circ \mathbf{w}_{2,t}).$$

The vectors $\mathbf{w}_{2,t}$ can be represented in the basis $\{\mathbf{e}_1, \ldots, \mathbf{e}_{M_2}\}$. Utilizing the orthonormality of the basis:

$$y_{2,t} = \sum_{m=1}^{M_2} (\mathbf{x} \circ \mathbf{w}_{1,m}) \cdot w_{2,t,m} = \sum_{m=1}^{M_2} \mathbf{x} \circ (w_{2,t,m} \cdot \mathbf{w}_{1,m}).$$

The above equality means that for each input signal $\mathbf{x}$ the considered two-layer linear trained perceptron gives the same output signal $\mathbf{y}_2$ as the one-layer linear perceptron with the suitably selected weights. □

Corollaries 8.8, 9.1.2 and Lemma 9.1.3 imply directly the following corollary.

Corollary 9.1.4 *The function* F *which corresponds to the trained linear* $M-$*perceptron with input signals from the set* X *is a linear operator. The set* X *is its domain.*

9.2 Dynamics of Training Process of Linear Perceptrons

In Sect. 8.3 it was shown that the formula

$$\frac{d\mathbf{w}}{dt} = -grad\ E(\mathbf{w}),$$

where $\mathbf{w}$ is a vector of all weights in the perceptron,

$$E(\mathbf{w}) = \sum_{n=1}^{N} \sum_{t=1}^{T_R} \left[y(\mathbf{w})_{R,t}^{(n)} - z_t^{(n)} \right]^2,$$

and

$$y(\mathbf{w})_{R,t}^{(n)} = f_{R,t} \left(\sum_{m=1}^{M} x_m^{(n)} w_{R,t,m} \right)$$

describes the dynamics of the training process of a perceptron. Let us consider a single linear $M-$neuron. Let a training set be given. As it has been shown above, it can be assumed that it has identity as its activation function. In such a case the error function is of the form

$$E(\mathbf{w}) = E(w_1, \ldots, w_M) = \sum_{n=1}^{N} \left[y(w_1, \ldots, w_M)^{(n)} - z^{(n)} \right]^2,$$

where

$$y(\mathbf{w}) = y(w_1, \ldots, w_M)^{(n)} = \sum_{m=1}^{M} x_m^{(n)} w_m.$$

Let us calculate the partial derivative from the equality (8.2)

$$\frac{\partial E(w_1, \ldots, w_M)}{\partial w_{m_0}} = \frac{\partial}{\partial w_{m_0}} \sum_{n=1}^{N} \left[\sum_{m=1}^{M} x_m^{(n)} w_m - z^{(n)} \right]^2.$$

Let us denote

$$H^{(n)} = \sum_{m=1}^{M} x_m^{(n)} w_m - z^{(n)}.$$

Then

$$\frac{\partial E(w_1, \ldots, w_M)}{\partial w_{m_0}} = \sum_{n=1}^{N} 2 \cdot H^{(n)} \cdot \frac{\partial (y^{(n)} - z^{(n)})}{\partial w_{m_0}} = 2 \cdot \sum_{n=1}^{N} H^{(n)} \cdot \frac{\partial y^{(n)}}{\partial w_{m_0}} =$$

$$= 2 \cdot \sum_{n=1}^{N} H^{(n)} \cdot \frac{\partial (\sum_{m=1}^{M} x_m^{(n)} w_m)}{\partial w_{m_0}} = 2 \cdot \sum_{n=1}^{N} H^{(n)} \cdot x_{m_0}^{(n)} =$$

$$= 2 \cdot \sum_{n=1}^{N} x_{m_0}^{(n)} \cdot \left[\left(\sum_{m=1}^{M} x_m^{(n)} w_m \right) - z^{(n)} \right].$$

Thus, the formula that describes the training process of a linear neuron by using a training set $\{(\mathbf{x}^{(1)}, z^{(1)}), \ldots, (\mathbf{x}^{(N)}, z^{(N)})\}$ is a linear nonhomogeneous differential equation of the form

$$\frac{d\mathbf{w}}{dt} = \mathbf{A} \cdot \mathbf{w} - \mathbf{b}, \qquad (9.1)$$

where $\mathbf{A}$ is the following matrix

$$\mathbf{A} = -2\mathbf{G}(\overline{\mathbf{x}}_1, \ldots, \overline{\mathbf{x}}_M),$$

and $\overline{\mathbf{x}}_m$, $m = 1, \ldots, M$, denotes N-dimensional vector, with the components formed by the sequence of the mth component of the $\mathbf{x}$ elements of the training set, i.e. $\overline{\mathbf{x}}_m = \left[x_m^{(1)}, \ldots, x_m^{(N)} \right]$.

Thus, for an $M-$neuron and an $N-$element training set the Gram matrix $\mathbf{G}$ in equation (9.1) is a square $M \times M$ matrix which elements are given as

$$g_{ij} = \overline{\mathbf{x}}_i \circ \overline{\mathbf{x}}_j,$$

whereas the vector $\mathbf{b}$ is an M-dimensional vector with the components

$$b_m = \overline{\mathbf{x}}_m \circ \mathbf{z}, \quad \mathbf{z} = \left[z^{(1)}, \ldots, z^{(N)} \right].$$

Remarks The training process of a one-layer linear perceptron consisted of T neurons is described by a system of T mutually independent linear differential equations the same as the Eq. (9.1) which describes the training process of a single neuron. It can be shown in the following way.

$$\frac{\partial E}{\partial w_{t^*, m^*}} = \frac{\partial}{\partial w_{t^*, m^*}} \sum_{n=1}^{N} \sum_{t=1}^{T} \left[\left(\sum_{m=1}^{M} x_m^{(n)} w_{t,m} \right) - z_t^{(n)} \right]^2 =$$

$$= \frac{\partial}{\partial w_{t^*,m^*}} \left\{ \sum_{n=1}^{N} \left\{ \sum_{t=1, t \neq t^*}^{T} \left[\left(\sum_{m=1}^{M} x_m^{(n)} w_{t,m} \right) - z_t^{(n)} \right]^2 \right\} + \right.$$

$$+ \frac{\partial}{\partial w_{t^*,m^*}} \sum_{n=1}^{N} \left[\left(\sum_{m=1}^{M} x_m^{(n)} w_{t^*,m} \right) - z_t^{(n)} \right]^2.$$

Since the first term does not depend on w_{t^*,m^*}, it zeroes. Thus:

$$\frac{\partial E}{\partial w_{t^*,m^*}} = \frac{\partial}{\partial w_{t^*,m^*}} \sum_{n=1}^{N} \left[\left(\sum_{m=1}^{M} x_m^{(n)} w_{t^*,m} \right) - z_t^{(n)} \right]^2 =$$

$$= 2 \cdot \sum_{n=1}^{N} x_{m^*}^{(n)} \left[\left(\sum_{m=1}^{M} x_m^{(n)} w_{t^*,m} \right) - z_{t^*}^{(n)} \right].$$

In such a way a system of equations indexed by a parameter $t^* \in \{1, \ldots, T\}$ has been obtained. It can be summed up as the following corollary.

Corollary 9.2.1 *The training process of a one-layer linear* M*-perceptron, which consists of* T *neurons, is described by a system of linear nonhomogeneous differential equations*

$$\frac{d\mathbf{w}_t}{dt} = -2 \cdot (G(\overline{\mathbf{x}}_1, \ldots, \overline{\mathbf{x}}_M) \cdot \mathbf{w}_t - \mathbf{B}_t). \tag{9.2}$$

Vector $\mathbf{B}_t$ *has, for a fixed* $t = t^*$, *the following components*

$$b_{m,t^*} = \overline{\mathbf{x}}_m \circ \mathbf{z}_{t^*},$$

where $\mathbf{z}_{t^*} = (z_{t^*}^{(1)}, \ldots, z_{t^*}^{(N)})$.

Remarks 1. The obtained system of equations can be written in the matrix form

$$\frac{d\mathbf{W}}{dt} = \mathbf{A} \cdot \mathbf{W} - \mathbf{B}, \tag{9.3}$$

where $\mathbf{W}$ is the matrix, whose the t-th column is the vector $\mathbf{w}_t$, whereas $\mathbf{B}$ is a matrix whose elements are given as $b_{m,t} = \mathbf{x}_m \circ \mathbf{z}_t$.

2. By Lemma 9.1.3, only one-layer linear perceptrons can be considered without loss of generality. Thus, Eq. (9.3) describes the most general case of the training process of linear perceptrons.

Example Let us consider a simple example - the training process of a single 2-neuron with the training set which consists of three elements:

$$\left\{ \left(\left[x_1^{(1)}, x_2^{(1)} \right], z^{(1)} \right), \ \left(\left[x_1^{(2)}, x_2^{(2)} \right], z^{(2)} \right), \ \left(\left[(x_1^{(3)}, x_2^{(3)} \right], z^{(3)} \right) \right\}.$$

This means that

$$\mathbf{x}^{(n)} = \left[x_1^{(n)}, x_2^{(n)} \right],$$

$$\bar{\mathbf{x}}_m = \left[x_m^{(1)}, x_m^{(2)}, x_m^{(3)} \right]$$

and

$$\mathbf{z} = \left[z^{(1)}, z^{(2)}, z^{(3)} \right].$$

According to the obtained formulae

$$\frac{dw_1}{dt} = -2 \sum_{n=1}^{3} x_1^{(n)} \cdot \left((x_1^{(n)} w_1 + x_2^{(n)} w_2) - z^{(n)} \right) =$$

$$-2 \cdot (x_1^{(1)} x_1^{(1)} w_1 + x_1^{(1)} x_2^{(1)} w_2 - x_1^{(1)} z^{(1)} + x_1^{(2)} x_1^{(2)} w_1 + x_1^{(2)} x_2^{(2)} w_2 - x_1^{(2)} z^{(2)} +$$

$$+ x_1^{(3)} x_1^{(3)} w_1 + x_1^{(3)} x_2^{(3)} w_2 - x_1^{(3)} z^{(3)}) =$$

$$= -2 \cdot ((x_1^{(1)} x_1^{(1)} + x_1^{(2)} x_1^{(2)} + x_1^{(3)} x_1^{(3)}) w_1 +$$

$$+ (x_1^{(1)} x_2^{(1)} + x_1^{(2)} x_2^{(2)} + x_1^{(3)} x_2^{(3)}) w_2 -$$

$$- x_1^{(1)} z^{(1)} - x_1^{(2)} z^{(2)} - x_1^{(3)} z^{(3)}) =$$

$$= -2 \cdot ((\bar{\mathbf{x}}_1 \circ \bar{\mathbf{x}}_1) \cdot w_1 + (\bar{\mathbf{x}}_1 \circ \bar{\mathbf{x}}_2) \cdot w_2 - (\bar{\mathbf{x}}_1 \circ \mathbf{z})).$$

The term $\frac{dw_2}{dt}$ can be calculated in the same way. To sum up, the following system of equations has been obtained.

$$\frac{dw_1}{dt} = -2 \cdot ((\bar{\mathbf{x}}_1 \circ \bar{\mathbf{x}}_1) \cdot w_1 + (\bar{\mathbf{x}}_1 \circ \bar{\mathbf{x}}_2) \cdot w_2 - (\bar{\mathbf{x}}_1 \circ \mathbf{z}))$$

$$\frac{dw_2}{dt} = -2 \cdot ((\bar{\mathbf{x}}_2 \circ \bar{\mathbf{x}}_1) \cdot w_1 + (\bar{\mathbf{x}}_2 \circ \bar{\mathbf{x}}_2) \cdot w_2 - (\bar{\mathbf{x}}_2 \circ \mathbf{z})),$$

which can be written as

$$\frac{d\mathbf{w}}{dt} = -2 \cdot (\mathbf{A} \cdot \mathbf{w} - \mathbf{B}),$$

where

$$\mathbf{A} = \left(\begin{array}{cc} \bar{\mathbf{x}}_1 \circ \bar{\mathbf{x}}_1 & \bar{\mathbf{x}}_1 \circ \bar{\mathbf{x}}_2 \\ \bar{\mathbf{x}}_2 \circ \bar{\mathbf{x}}_1 & \bar{\mathbf{x}}_2 \circ \bar{\mathbf{x}}_2 \end{array} \right) \tag{9.4}$$

and

$$\mathbf{B} = \begin{pmatrix} \overline{\mathbf{x}}_1 \circ \mathbf{z} \\ \overline{\mathbf{x}}_2 \circ \mathbf{z} \end{pmatrix}$$

9.3 Stability of the Learning Process of Linear Perceptrons

Let us consider the stability of training process of linear perceptrons. Without loss of generality, one-layer linear perceptron with identity activation function can be considered - see Corollary 9.1.1 and Lemma 9.1.3. Furthermore, the learning process of a one-layer linear perceptron is described by a system of mutually independent linear differential equations - see formula (9.2). Therefore it is sufficient to analyze one equation that describes the training process of single linear neuron.

Theorem 9.3.1 *The flow generated by the nonhomogeneous linear differential equation*

$$\frac{d\mathbf{w}}{dt} = -2 \cdot \mathbf{G}(\overline{\mathbf{x}}_1, \ldots, \overline{\mathbf{x}}_M)\mathbf{w} - \mathbf{B} \tag{9.5}$$

is asymptotically stable if and only if the vectors $\{\overline{\mathbf{x}}_1, \ldots, \overline{\mathbf{x}}_M\}$ are linearly independent.

Proof By Lemma 5.3.1 it is sufficient to check asymptotic stability of the linear differential equation

$$\frac{d\mathbf{w}}{dt} = \mathbf{A} \cdot \mathbf{w},$$

where $\mathbf{A} = -2\mathbf{G}(\overline{\mathbf{x}}_1, \ldots, \overline{\mathbf{x}}_M)$. The stability can be proved by using Hurwitz criterion - see Theorem 5.3.5. Let us calculate all Δ_i and let us show that they all have positive values.

$$\Delta_1 = -A_1 = -(-2) \cdot Tr\, \mathbf{G}(\overline{\mathbf{x}}_1, \ldots, \overline{\mathbf{x}}_M) = 2 \cdot \sum_{m=1}^{M} \overline{\mathbf{x}}_m \circ \overline{\mathbf{x}}_m > 0,$$

It can be easily shown that Δ_2 is positive, as well.

$$\Delta_2 = -A_1 \cdot A_2 + A_3 > 0.$$

The above condition is equivalent to the following one

$$A_3 > A_1 \cdot A_2, \quad \text{or} \quad (-2)^3 \cdot G_3 > (-2)^3 \cdot G_1 \cdot G_2,$$

where G_k is a sum of all principal minors of the rank k of the matrix G.

Thus, it has been obtained

$$G_3 < G_1 \cdot G_2. \tag{9.6}$$

On the left side of (9.6) the sum of all principal minors of rank 3 occurs. It can be observed that for each component of the left side of inequality (9.6) there exists a component on the right side such that Corollary 4.12 can be applied. Furthermore, on the right side there exist additional components whose positiveness is directly implied by the linear independence of the vectors of the matrix G. Thus, inequality (9.6) is satisfied.

By Corollary 4.10, the remaining assumptions of Hurwitz criterion:

$$\Delta_n = (-1)^n \cdot A_n \cdot \Delta_{n-1} = (-1)^n \cdot (-2)^n \cdot G_n \cdot \Delta_{n-1} > 0,$$

for $n \geq 3$ are satisfied if and only if the vectors $\{\overline{\mathbf{x}}_1, \dots, \overline{\mathbf{x}}_M\}$ are linearly independent.
□

Thus, it has been proved that the dynamics of the flow generated by equation (9.5) is asymptotically stable and, as a consequence, by Corollary 5.3.3, it is globally asymptotically stable as well provided that the vectors $\overline{\mathbf{x}}_1, \dots, \overline{\mathbf{x}}_M$ are linearly independent. Let us recall that if the vectors have N components and $N \geq M$ then linear independency is a typical property - see Lemma 4.11. This implies the following Theorem.

Theorem 9.3.2 *The time continuous model (9.2) of the training process of the linear perceptron that consists of M-neurons is, generically, globally asymptotically stable if only $N \geq M$, where N is the length of training set. Furthermore, the flow generated by differential equation (9.5) that describes the training process has exactly one fixed point which is hyperbolic, globally attracting.*

The above theorem implies that the flow which models the training process of a linear perceptron has, generically, extremely regular dynamics. Nevertheless, the question arises, whether this regularity is preserved if a numerical method is applied in the training process implementation. It turns out that the answer is affirmative.

First of all, the dynamics of discretization of the flow generated by the system Eq. (9.2) has the same dynamics as the Euler method applied to (9.2).

Theorem 9.3.3 *Let us assume that the linear perceptron consists of $M-$ neurons and that the training set consists of N elements, $N \geq M$. The discretization Φ_h of the flow Φ generated by (9.2) is, for a sufficiently small h, generically topologically conjugate with the cascade Ψ_h generated by the Euler method applied to (9.2).*

Proof Using Fečkan theorem is the simplest way to prove this theorem. Since the equations are mutually independent, it is sufficient to consider only one equation that describes training process of an $M-$neuron.

Put $g \equiv 0$ in (5.8). Since the flow $\Phi : \mathbb{R}^M \rightarrow \mathbb{R}^M$ has, generically, one hyperbolic attracting fixed point, the assumptions of Fečkan theorem are satisfied. Thus, for sufficiently small h, there exists a homeomorphism α_h^* that conjugates the cascades Φ_h and Ψ_h on a ball B_r that is a neighbourhood of the fixed point. Since the attracting fixed point is global, for each $\mathbf{x} \in \mathbb{R}^M$ which does not belong to B_r there exists $n_\mathbf{x}$ such that $\Phi_h^{-n_\mathbf{x}} \in B_r$ and $\Phi_h^{-n_\mathbf{x}-1} \notin B_r$. Let us define

$$\alpha_h(\mathbf{x}) := \begin{cases} \alpha_h^*(\mathbf{x}) & \text{if } \mathbf{x} \in B_r, \\ \left(\Phi_h^{-n_\mathbf{x}} \circ \alpha_h^* \circ \Psi_h^{n_\mathbf{x}} \right)(\mathbf{x}) & \text{otherwise.} \end{cases} \tag{9.7}$$

The way of constructing α_h is, so called, a basic domain method. It is a commonly known fact that the the mapping α_h defined in such a way is a conjugating homeomorphism.

As it has been aforementioned, the numerical computer implementations can be done in the approximate arithmetic, not in the exact one. The dynamical properties are preserved if the system has shadowing property.

Theorem 9.3.4 *Let $\mathcal{T} = \Theta_c$ or $\mathcal{T} = \Theta_s$. Under assumptions of Theorem 9.3.3 the cascade Φ_h is, for a sufficiently small h, generically $\mathcal{T}$ robust.*

Proof Generically, the cascade Φ_h has one global hyperbolic attracting fixed point. This implies that it is a gradient-like Morse-Smale cascade. By Theorem 5.7.8 it is $\mathcal{T}$ robust. □

9.4   Bibliographic Remarks

The results described in this chapter are presented in [32].

Chapter 10
Weakly Nonlinear Perceptrons

The character of the dynamics of linear both flows and cascades is well investigated. It is known, among others, that the cascades generated by a linear flow preserve regular dynamics of the flow. In particular, the problems connected with topological conjugacy and shadowing properties are resolved. The regular dynamics of linear dynamical systems is preserved if they are weakly distorted - so called weakly nonlinear dynamical systems. This is formulated strictly in Grobman and Hartman theorems and Fečkan Theorem - see Sect. 5.6.

The results obtained for weakly nonlinear dynamical systems can be applied for perceptrons. Basing on the aforementioned class of dynamical systems a new type of perceptrons that consist of so called weakly nonlinear neurons has been introduced. It turns out that this type of neurons preserves regular dynamics of training process of linear perceptrons. Furthermore, they have stronger approximation properties than linear perceptrons. Thus, there exists a class of problems that can be solved by using weakly nonlinear neurons and they cannot be solved by using linear neurons.

Definition 10.1 A neuron is called a weakly nonlinear neuron if its activation function is of the form

$$f : \mathbb{R} \ni s \rightarrow f(s) = s + \tilde{g}(s) \in \mathbb{R},$$

where $\tilde{g} : \mathbb{R} \rightarrow \mathbb{R}$ is bounded and, furthermore, $|\tilde{g}'(x)| < b_1$ and $|\tilde{g}''(x)| < b_2$ for each x, provided that the constants $b_1 > 0$ and $b_2 > 0$ are sufficiently small.

It can be shown that the training process of a weakly nonlinear neuron and, as the consequence, a one-layer weakly nonlinear perceptron, is modelled by a flow generated by differential equation 5.8 which satisfies the assumptions of Theorem 5.6.7.

© Springer International Publishing AG, part of Springer Nature 2019
A. Bielecki, *Models of Neurons and Perceptrons: Selected Problems and Challenges*, Studies in Computational Intelligence 770,
https://doi.org/10.1007/978-3-319-90140-4_10

Theorem 10.2 *The flow model of the training process of a weakly nonlinear neuron with a square error function (8.4) has the same dynamics as the flow generated by the differential equation (5.8) which satisfies the assumptions of Theorem 5.6.7.*

Proof Let us assume that an $N-$elementary training set $\left((\mathbf{x}^{(1)}, z^{(1)}), \ldots, (\mathbf{x}^{(N)}, z^{(N)})\right)$ of an $M-$neuron is given. The total excitation of the neuron in the $n-$th training step is of the form

$$s^{(n)} = \sum_{m=1}^{M} x_m^{(n)} w_m^{(n)}.$$

Since the considered neuron is weakly nonlinear, its activation function has the form

$$f\left(s^{(n)}\right) = s^{(n)} + \tilde{g}\left(s^{(n)}\right),$$

whereas the square error function is as follows

$$E(w_1, \ldots, w_M) = \frac{1}{2} \sum_{n=1}^{N} \left[s^{(n)} + \tilde{g}\left(s^{(n)}\right) - z^{(n)}\right]^2.$$

Let us calculate the $k-$th component of the gardient

$$\frac{\partial E}{\partial w_k} = \sum_{n=1}^{N} \left\{ \left[s^{(n)} + \tilde{g}(s^{(n)}) - z^{(n)}\right] \cdot \left[x_k^{(n)} + \tilde{g}'(s^{(n)}) \cdot x_k^{(n)}\right] \right\} =$$

$$= \sum_{n=1}^{N} x_k^{(n)} s^{(n)} + \sum_{n=1}^{N} x_k^{(n)} \cdot \left[\tilde{g}(s^{(n)}) + \tilde{g}'(s^{(n)}) \cdot f(s^{(n)}) - z^{(n)} \cdot \tilde{g}'(s^{(n)}) - z^{(n)}\right].$$

Thus, the gradient differential equation that models the training process of a weakly nonlinear neuron

$$\frac{d\mathbf{w}}{dt} = -grad\ E(\mathbf{w})$$

has the following form

$$\frac{d\mathbf{w}}{dt} = -(\mathbf{A}\mathbf{w} + \mathbf{b} + \mathbf{g}(\mathbf{w})), \tag{10.1}$$

where the Gram matrix $\mathbf{A}$ is given by the formula

$$\mathbf{A} = \begin{pmatrix} \overline{\mathbf{x}}_1 \circ \overline{\mathbf{x}}_1 & \cdots & \overline{\mathbf{x}}_1 \circ \overline{\mathbf{x}}_M \\ \vdots & & \vdots \\ \overline{\mathbf{x}}_M \circ \overline{\mathbf{x}}_1 & \cdots & \overline{\mathbf{x}}_M \circ \overline{\mathbf{x}}_M \end{pmatrix}, \tag{10.2}$$

and the vector $\mathbf{b}$ has the form

$$\mathbf{b} = - \begin{pmatrix} \sum_{n=1}^{N} x_1^{(n)} z^{(n)} \\ \vdots \\ \sum_{n=1}^{N} x_M^{(n)} z^{(n)} \end{pmatrix}, \tag{10.3}$$

whereas the vector mapping $\mathbf{g} : \mathbb{R}^M \to \mathbb{R}^M$ is of the form

$$\mathbf{g}(\mathbf{w}) = \begin{pmatrix} \sum_{n=1}^{N} x_1^{(n)} \cdot \left[\tilde{g}(s^{(n)}) + \tilde{g}'(s^{(n)}) \cdot f(s^{(n)}) - z^{(n)} \cdot \tilde{g}'(s^{(n)}) \right] \\ \vdots \\ \sum_{n=1}^{N} x_M^{(n)} \cdot \left[\tilde{g}(s^{(n)}) + \tilde{g}'(s^{(n)}) \cdot f(s^{(n)}) - z^{(n)} \cdot \tilde{g}'(s^{(n)}) \right] \end{pmatrix}.$$

Let us calculate the element of the $D\mathbf{g}(\mathbf{w})$ matrix

$$D\mathbf{g}(\mathbf{w})_{ik} = \frac{\partial}{\partial w_i} \left(\sum_{n=1}^{N} x_k^{(n)} \cdot \left[\tilde{g}(s^{(n)}) + \tilde{g}'(s^{(n)}) \cdot f(s^{(n)}) - z^{(n)} \cdot \tilde{g}'(s^{(n)}) \right] \right) =$$

$$= \sum_{n=1}^{N} x_k^{(n)} \cdot \left[\frac{\partial}{\partial w_i} \tilde{g}(s^{(n)}) + \frac{\partial}{\partial w_i} (\tilde{g}'(s^{(n)}) \cdot f(s^{(n)})) - z^{(n)} \cdot \frac{\partial}{\partial w_i} \tilde{g}'(s^{(n)}) \right] =$$

$$= \sum_{n=1}^{N} x_k^{(n)} \cdot \left[\frac{\partial}{\partial w_i} \tilde{g}(s^{(n)}) + f(s^{(n)}) \cdot \frac{\partial}{\partial w_i} \tilde{g}'(s^{(n)}) + \tilde{g}'(s^{(n)}) \cdot \frac{\partial}{\partial w_i} f(s^{(n)}) - z^{(n)} \cdot \frac{\partial}{\partial w_i} \tilde{g}'(s^{(n)}) \right] =$$

$$= \sum_{n=1}^{N} x_k^{(n)} \cdot \left[x_i^{(n)} \tilde{g}'(s^{(n)}) + f(s^{(n)}) \cdot x_i^{(n)} \tilde{g}''(s^{(n)}) + \tilde{g}'(s^{(n)}) \cdot x_i^{(n)} f'(s^{(n)}) - z^{(n)} \cdot x_i^{(n)} \tilde{g}''(s^{(n)}) \right] =$$

$$= \sum_{n=1}^{N} x_k^{(n)} x_i^{(n)} \cdot \left[\tilde{g}'(s^{(n)}) + f(s^{(n)}) \cdot \tilde{g}''(s^{(n)}) + \tilde{g}'(s^{(n)}) \cdot f'(s^{(n)}) - z^{(n)} \cdot \tilde{g}''(s^{(n)}) \right].$$

Let us write Eq. (10.1) as follows

$$\frac{d\mathbf{w}}{dt} = -\mathbf{A}(\mathbf{w} + \mathbf{A}^{-1}\mathbf{b} + \mathbf{A}^{-1}\mathbf{g}(\mathbf{w}))), \tag{10.4}$$

and let us put $\mathbf{z} = \mathbf{w} + \mathbf{A}^{-1}\mathbf{b}$. This substitution defines topological conjugacy and the Eq. (10.4) is transformed into the following form

$$\frac{d\mathbf{z}}{dt} = -(\mathbf{A}\mathbf{z} + \mathbf{g}(\mathbf{z} - \mathbf{A}^{-1}\mathbf{b})), \tag{10.5}$$

If $\mathbf{g}(\mathbf{A}^{-1}\mathbf{b}) = 0$ then the mapping $\mathbf{g}$ satisfies the assumptions of Fečkan theorem. □

To sum up, if only the modules of the mapping $\mathbf{g}$ and its first and second differential satisfy the conditions specified as the assumptions of Theorem 10.2, then the mapping satisfies the assumptions of both Grobman–Hartman and Fečkan Theorems. This means that

- The flow generated by the equation

$$\frac{d\mathbf{w}}{dt} = -(\mathbf{Aw} + \mathbf{b} + \mathbf{g}(\mathbf{w}))$$

  is globally topologically conjugate with its linear part

$$\frac{d\mathbf{w}}{dt} = -\mathbf{Aw}.$$

  This implies, among others, the asymptotic stability of the training process.
- The cascade generated by the discretization of the equation

$$\frac{d\mathbf{w}}{dt} = -(\mathbf{Aw} + \mathbf{b} + \mathbf{g}(\mathbf{w}))$$

  is, on a large ball, topologically conjugate with the cascade generated by the Euler method for this equation. It should be stressed that not the discretization but the numerical method is the basis of the training algorithm of a neural network. Topological conjugacy on a large ball is, in practice, sufficient - see the discussion in the next section.
- A linear flow that models the training process of a linear perceptron is, generically, globally asymptotically stable with an attracting fixed point - see Theorem 9.3.2, stable under numerics - Theorem 9.3.3 and T-robust, where $T = \theta_c \cup \theta_s$ - see Theorem 9.3.4. By Theorem 10.2 and the fact that these properties are invariant under topological conjugacy the training process of a weakly nonlinear perceptron has the aforementioned properties, as well.

10.1   Bibliographic Remarks

The idea of a weakly nonlinear perceptron was proposed by the author [29] and it was studied by him [32, 40].

Chapter 11
Nonlinear Perceptrons

In this chapter a training process of the most general class of perceptrons - the nonlinear ones - is considered. Runge–Kutta methods, first of all the gradient descent method (the Euler method), that are used as the numerical training algorithm, are studied in the context of their stability and robustness. It should be stressed that the continuous model of the training process is considered in the Euclidean space $\mathbb{R}^n$. The training algorithm is implemented as an iterative numerical rule in $\mathbb{R}^n$, as well. However, the theoretical analysis presented in this chapter concerns numerical schemata on the $n-$dimensional compact manifold $\mathcal{M}_S^n$, which is homeomorphic to the sphere $\mathcal{S}^n$. This is possible thanks to the specific compactification procedure, which is described in details in the Step 1 of the proof of Theorem 11.1. Such approach allows us to apply results concerning numerical dynamics on compact manifolds.

First of all let us notice that most types of the activation function used in practice are of the class $C^2(\mathbb{R}, \mathbb{R})$. Bipolar and unipolar sigmoid functions and most radial functions, satisfy this assumption. Therefore, the square error function E is of the class $C^2(\mathbb{R}^n, \mathbb{R})$, as well. Equation (8.1) describes an algorithm of finding a local minimum of the error function E by using the descent gradient method (the Euler method) which is the Runge–Kutta method of order $k = 1$. It should be mentioned that the Runge–Kutta methods of orders $k = 2$ are also applied as training algorithms of perceptrons.

In order to applied the aforementioned compactification, a regularization procedure of the error function, which in gradient equation plays a role of the potential, has to be applied. Let $B^n(\mathbf{0}, r)$ denotes a closed, n-dimensional ball in $\mathbb{R}^n$, where $\mathbf{0}$ denotes zero in $\mathbb{R}^n$. Theorem 11.1 can be applied to the dynamical systems on compact manifold within boundaries. In order to apply the theorem to perceptrons training process, the process must be transformed onto such manifold. In order to perform this transformation let us modify the error function in such a way that on a certain, sufficiently large ball $B^n(\mathbf{0}, r_1) \subset \mathbb{R}^n$, the potential is not modified. The radius r_1 can be choose as large as it is needed. The potential will be modified in

© Springer International Publishing AG, part of Springer Nature 2019
A. Bielecki, *Models of Neurons and Perceptrons: Selected Problems
and Challenges*, Studies in Computational Intelligence 770,
https://doi.org/10.1007/978-3-319-90140-4_11

such a way that the ball $B^n(\mathbf{0}, 2r_1)$ will be an invariant set of the considered dynamical system which models the training process. Thus, let $E(\mathbf{w}) = E(r)$, $r = \|\mathbf{w}\|^2$, for r large but less than $2r_1$. In such a way a flow $\left(B^n(\mathbf{0}, 2r_1), \tilde{\phi}\right)$ is obtained. The procedure of the regularization of the error function E is presented in details in the second step of the proof.

Let us notice that the modification of the error function consists in remaining the potential unchanged in a large ball and modification outside the ball according to needs. A such way of modification is well based on the properties on the modelled realities. If an algorithm is implemented on the computer, then the range of represented numbers is bounded. Therefore, it can be assumed, without loss of generality, that modules of the considered vectors are less than r_1. Let us also mentioned that the situation in neurodynamics is similar. In a biological neural cell, neurotransmitters are liberated in tiny amounts from the presynaptic bouton to synaptic cleft - about 10^{-17} mol per impulse.

Thus, both in artificial and biological neural networks, norms of input vectors $\mathbf{x}$ and weight vectors $\mathbf{w}$ are bounded. Therefore, in models of neurons, without loss of generality, only bounded vectors can be considered. This means, among others, that only dynamics restricted to some set, possibly large but bounded, is essential. We can assume that this set is a ball $B^n(\mathbf{0}, r_1)$ with the radius r_1 sufficiently large.

To sum up, the error function remains unchanged on the ball $B^n(\mathbf{0}, r_1)$ and the flow $\left(B^n(\mathbf{0}, 2r_1), \tilde{\phi}\right)$ which is generated by the differential equation

$$\frac{d\mathbf{w}}{dt} = -grad\ \tilde{E}(\mathbf{w}),\ \mathbf{w} \in B^n(\mathbf{0}, R) \subset \mathbb{R}^n \tag{11.1}$$

is adequate as the model of a perceptron training process.

Let Γ denotes the set of all C^1 vector fields on the manifold $\mathcal{M}$. Let us assume that the vector fields are equipped with the C^1 topology. Let us also assume that $\mathcal{G} \subset \Gamma$ is formed by all vector fields that have the form $-grad\ E$, provided that $E : \mathcal{M} \to \mathbb{R}$ is a function of the class C^2. With any vector field in $\mathcal{G}$ let two cascades be associated: its discretizations: ϕ_T and Runge–Kutta methods $\psi_{\frac{T}{m}, p}$. Let us furthermore assume that $\Psi = \psi_{\frac{T}{m}, p}^m$ - see Theorem 5.6.11 and $T = \Theta$ - see Sect. 5.7. The dynamical properties of a training process of a perceptron which has n weights can be specified in the form of the following theorem.

Theorem 11.1 *Let a real number $T > 0$ be given. Assume that a training process of a perceptron which has n weights is modelled by a flow $\tilde{\phi}$ on the ball $B^n(\mathbf{0}, 2r_1) \subset \mathbb{R}^n$ - see formula (11.1). Then, there exists a compact, smooth, n−dimensional manifold $\mathcal{M}_S^n$ without boundaries and a flow $\left(\mathcal{M}_S^n, \widehat{\phi}\right)$ such that $B^n(\mathbf{0}, 2r_1) \subset \mathcal{M}_S^n$, $\left(\mathcal{M}_S^n | B^n(\mathbf{0}, 2r_1), \widehat{\phi}\right) = \left(B^n(\mathbf{0}, 2r_1), \tilde{\phi}\right)$. Furthermore, the flow $\left(\mathcal{M}_S^n, \widehat{\phi}\right)$ is, generically, stable under numerics with respect to the operator Ψ i.e. the cascades $\left(\mathcal{M}_S^n, \widehat{\phi}_T\right)$ and $\left(\mathcal{M}_S^n, \Psi\right)$ are topologically conjugate. Moreover, the cascades $\widehat{\phi}_T$ and Ψ are generically T robust.*

Proof **Step 1. Construction of the manifold $\mathcal{M}^n_S$.**

Let $B^n(\mathbf{0}, r_1) \subset \mathbb{R}^n$ be a closed ball which has the radius r_1 as large as we need and put $R = 3r_1$. A manifold $\mathcal{M}^n_S \in \mathbb{R}^n \times \mathbb{R}$ will be constructed in such a way that it will have a radial symmetry with respect to rotations around the real axis which is orthogonal to the $n-$ dimensional Euclidean hyperplane Euc^n - see Fig. 11.1. Let us assume that the ball $B^n(\mathbf{0}, R)$ is contained in Euc^n. Since the manifold has the radial symmetry, its construction can be described for $n-$ dimensional section. Let us describe it for $n = 2$. The construction can be generalised without any problems for higher dimensions. Thus, let us glue the line segment $[-R, R]$, which is contained in Euc^2, with two hemicircles of a circle of radius r_s at the points $A = (R, 0)$ and $D = (-R, 0)$ respectively. Then, let us glue the obtained curve with the line segment parallel to the previous one at the points $C = (-R, r_s)$ and $D = (R, r_s)$, respectively. The construction is illustrated in Fig. 11.1. The obtained manifold is compact because it is homeomorphic to 2−dimensional sphere S^2. Furthermore, it is of the class $\mathcal{C}^1$. The lack of $\mathcal{C}^\infty$ smoothness in the points A, B, C, D on a two-dimensional section can be counterbalanced by using a mollifier function $f_{[a,b]}(x) \in \mathcal{C}^\infty(\mathbb{R})$. Let $f_{[a,b]}$ be defined as: $f_{[a,b]}(x) = 0$ for $x \in (-\infty, a]$, $f_{[a,b]}(x) = 1$ for $x \in [b, \infty)$ and $f_{[a,b]}$ is increasing on $[a, b]$. Such type of function is called a cutoff function and is commonly used. Thanks to the symmetry of the two-dimensional section, it is sufficient to describe the smoothing procedure only at the point A. We can treat the quarter of the section as the function $f_{sec} : [0, R + r_s] \to [0, r_s]$ defined in the following form:

$$f_{sec}(x) := \begin{cases} 0 & \text{for } r \in [0, R), \\ r_s - \sqrt{r_s^2 - (x - R)^2} & \text{for } [R, R + r_s]. \end{cases}$$

Thus, the point $A = (R, 0)$ is a glue point. Let us cut the domain of $f_{[R,R+\frac{r_s}{2}]}$ to the interval $[0, R + r_s]$. Define the mapping $f_{smooth}(x) := f_{sec}(x) \cdot f_{[R,R+\frac{r_s}{2}]}(x)$. It is of the class $\mathcal{C}^\infty(0, R + r_s)$. The manifold $\mathcal{M}^2_S$ is obtained by the rotation of

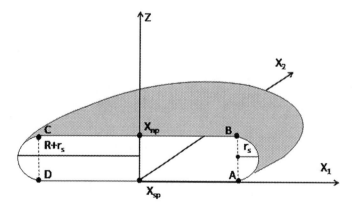

Fig. 11.1 Construction of the manifold $\mathcal{M}^n_S$ for $n = 2$.

the described two-dimensional section glued smoothly at the points A, B, C and D around the real axis - see Fig. 11.1.

Let $Base$ denotes the part of $\mathcal{M}_S^n$ which belongs to the hyperplane Euc^n, i.e. $Base := B^n(\mathbf{0}, R)$. Define $Cap := \mathcal{M}_S^n \setminus Base$. Let us notice that the flow $\tilde{\phi}$ is founded on the ball $B^n(\mathbf{0}, 2r_1) \subset Euc^n$ which is a subset of $Base$.

Step 2. Compactification.

As it has been aforementioned, in computer implementations vectors are bounded. Therefore, only the training process on a closed, sufficiently large ball $B^n(\mathbf{0}, r_1)$ is essential. Thus, modifications and completing of the potential outside the ball $B^n(\mathbf{0}, r_1)$ can be done only if the ball remains invariable. Let us modify the potential E by using a function g defined in the following way

$$g(\mathbf{w}) := \begin{cases} 1 & \text{for } r \in [0, r_1), \\ e^{(r-r_1)^a} & \text{for } r \geq r_1, \end{cases}$$

where $r := \|\mathbf{w}\|^2$. The square dependence is chosen for clarity because then $\frac{\partial r}{\partial w_i} = 2w_i$ provided that $\| \cdot \|$ is the Euclidean norm. The natural number a is selected in dependence on the potential E. The radius r_1 is chosen in the way specified below. Let us notice that $g \in C^2(\mathbb{R}^n, \mathbb{R})$ for $a > 2$. Define $\tilde{E}(\mathbf{w}) := E(\mathbf{w}) \cdot g(\mathbf{w})$. If a is sufficiently large, then the solutions of the equation $\mathbf{\dot{w}} = -grad\, \tilde{E}(\mathbf{w})$, which generates a flow $\tilde{\phi}$, cut the (n-1)-dimensional sphere $S^{n-1}(\mathbf{0}, 2r_1) \subset Base$ transversally entering into interior of the ball $B^n(\mathbf{0}, 2r_1)$. This is equivalent to the fact that the scalar product $-grad\, \tilde{E}(\mathbf{w}) \circ \mathbf{w}$ has for $r = 2r_1$ negative values. As a consequence, the ball $B^n(\mathbf{0}, 2r_1)$ is an invariant set of the flow $\tilde{\phi}$. To show this, let us calculate the ith component of the scalar product $-grad\, \tilde{E}(\mathbf{w}) \circ \mathbf{w}$

$$-w_i \cdot \frac{\partial \tilde{E}(\mathbf{w})}{\partial w_i} = -w_i \cdot \frac{\partial}{\partial w_i}(E(\mathbf{w}) \cdot g(\mathbf{w})) =$$

$$= -w_i \left(E(\mathbf{w}) \cdot \frac{\partial g(\mathbf{w})}{\partial w_i} + g(\mathbf{w}) \cdot \frac{\partial E(\mathbf{w})}{\partial w_i} \right) = \dots$$

Let us calculate the derivative

$$\frac{\partial g(\mathbf{w})}{\partial w_i} = \begin{cases} 2 \cdot w_i \cdot a \cdot (r - r_1)^{a-1} \cdot e^{(r-r_1)^a} & \text{for } r > r_1, \\ 0 & \text{for } r \in [0, r_1], \end{cases}$$

Let us put $r := 2r_1$. Continuing the calculation, we obtain

$$\dots = \left(-2w_i^2 a(r - r_1)^{a-1} E(\mathbf{w}) - w_i \frac{\partial E(\mathbf{w})}{\partial w_i} \right) e^{(r-r_1)^a} =$$

$$\left(-2w_i^2 ar_1^{a-1} E(\mathbf{w}) - w_i \frac{\partial E(\mathbf{w})}{\partial w_i} \right) e^{r_1^a}.$$

Thus, as on $\mathcal{S}^{n-1}(\mathbf{0}, 2r_1)$ we have $\sum_i w_i^2 := \|\mathbf{w}\|^2 = 2r_1$, so

$$-grad\,\tilde{E}(\mathbf{w}) \circ \mathbf{w} = e^{r_1^a}\left(-4ar_1^a E(\mathbf{w}) - \sum_i w_i \frac{\partial E(\mathbf{w})}{\partial w_i}\right).$$

The problem is considered on the sphere $\mathcal{S}^{n-1}(\mathbf{0}, 2r_1)$. Therefore, all variables, functions and derivatives are bounded. In particular, the term $-\sum_i w_i \frac{\partial E(\mathbf{w})}{\partial w_i}$ can have positive value but is upper bounded. The potential E is, by definition, nonnegative and the considered flow, as a gradient morse-Smale flow, has a finite number of singularities. This implies that E has only a finite number of zeroes. Therefore r_1 can be chosen in such a way that $E(\mathbf{w}) > 0$ for each $\mathbf{w}$ such that $\|\mathbf{w}\|^2 = 2r_1 > 0$. Since r_1 is large and the term $\sum_i w_i \frac{\partial E(\mathbf{w})}{\partial w_i}$ does not depend on a, the number a can be chosen so large that the inequality $4ar_1^a E(\mathbf{w}) > \left|\sum_i w_i \frac{\partial E(\mathbf{w})}{\partial w_i}\right|$ is satisfied. Let us restrict the domain of $\tilde{E}$ to $B^n(\mathbf{0}, 2r_1) \subset Base$. Complete the potential on $\mathcal{M}_S^n$ in such a way that at the point x_{np}, which corresponds to the north pole of the sphere (see Fig. 11.1), there exists a hyperbolic fixed point, repelling on $\mathcal{M}_S^n \setminus B^n(\mathbf{0}, 2r_1)$. Then, let us glue C^2-regularly the potential on the border of $B^n(\mathbf{0}, 2r_1)$. This can be done in the following way. Let us define the potential on $Cap \cup \partial Base$ as $V(\mathbf{w}) := c \cdot \varrho(x_{sp}, \mathbf{w})$, where $c > 0$ is chosen in such a way that the minimal value of V on the border of $Base$ is greater than the maximal value of $\tilde{E}$ on $B^n(\mathbf{0}, 2r_1)$. Let us assume that the points x_{sp} and x_{np} correspond to the south pole and to the north pole of the sphere, respectively. Let us also define a cutoff function g_γ, on a geodesic line γ, such that $g_\gamma(\mathbf{w}) = \tilde{E}(\gamma \cap B^n(\mathbf{0}, 2r_1))$ if $\varrho(x_{sp}, \mathbf{w}) \le 2r_1$ and $g_\gamma(\mathbf{w}) = V(\gamma \cap \partial Cap)$ if $\varrho(x_{sp}, \mathbf{w}) \ge R = 3r_1$. The geodesic line γ connects the points x_{sp} and x_{np}. Let us define

$$\widehat{E}(\mathbf{w}) := \begin{cases} \tilde{E}(\mathbf{w}) & \text{on } \operatorname{int} B^n(\mathbf{0}, 2r_1) \\ g_\gamma(\mathbf{w}) & \text{on } Base \setminus \operatorname{int} B^n(\mathbf{0}, 2r_1) \\ V(\mathbf{w}) & \text{on } Cap. \end{cases}$$

Thus, the potential $\widehat{E} \in C^2(\mathcal{M}_S^n)$ has been obtained. As a consequence, the dynamical system $(\mathcal{M}_S^n, \widehat{\phi})$, which is generated by differential equation $\mathcal{M}_S^n$

$$\frac{d\mathbf{w}}{dt} = -\operatorname{grad}\widehat{E}(\mathbf{w}) \tag{11.2}$$

has been constructed. Let us fix a time step T. By applying a Runge–Kutta method, the cascades $\left(B^n(\mathbf{0}, 2r_1), \tilde{\phi}_T\right)$, $\left(B^n(\mathbf{0}, 2r_1), \tilde{\psi}_{\frac{T}{m}}\right)$, $\left(\mathcal{M}_S^n, \widehat{\phi}_T\right)$ and $\left(\mathcal{M}_S^n, \widehat{\psi}_{\frac{T}{m}}\right)$ are generated. By the properties of $-grad\,\tilde{E}(\mathbf{w})$, the ball $B^n(\mathbf{0}, 2r_1)$ is the invariant set of the cascade $\tilde{\phi}_T$. Furthermore, for a sufficiently large m, it is also the invariant set of the cascade $\tilde{\psi}_{\frac{T}{m}}$. This implies that $B^n(\mathbf{0}, 2r_1)$ is invariant for $\widehat{\phi}_T$ and $\widehat{\psi}_{\frac{T}{m}}$ as well.

Step 3. Genericity.

Structural stability of a dynamical system is equivalent to the strong transversality condition and Axiom A - see [148], p. 171. Furthermore, if a dynamical system satisfies Axiom A, then it has only a finite number of singularities and all of them are hyperbolic. Moreover, the strong transversality condition implies that the gradient system has no saddle-saddle connections. Therefore, the structural stability of the flow $(\mathcal{M}_S^n, \widehat{\phi})$ that describes the training process of perceptrons implies the assumptions of Theorem 5.6.11. Thus, the assumptions of Theorem 5.6.11 are generic because the set of structurally stable systems is open and dense in the space of gradient dynamical systems $\mathcal{G}$ (see [148], p. 116).

Step 4. Stability under numerics.

The flow which models a training process of perceptrons has, inside the ball $B^n(\mathbf{0}, R) \subset Euc^n$, a finite number of singularities and all of them are hyperbolic ones. Therefore, after the compactification and modification of the potential, the modified flow (11.2) and, as a consequence, the cascades $\left(\mathcal{M}_S^n, \widehat{\phi}_T\right)$ and $\left(\mathcal{M}_S^n, \widehat{\psi}_{\frac{T}{m},p}\right)$ generated by this flow, constructed on the manifold $\mathcal{M}_S^n$, satisfy assumptions of Theorem 5.6.11. Thus, a training process of perceptron is, after compactification, generically stable under numerics with respect to the operator $\Psi = \widehat{\psi}_{\frac{T}{m},p}^m$ according to every Runge–Kutta method $\widehat{\psi}_{\frac{T}{m},p}$.

Step 5. Robustness.

Let us prove that a generic training process of a perceptron is robust. That means, by definition, thai it is shadowing and inverse shadowing with respect to a broad class of δ-methods. In order to do this, it will be shown that the sort of differential equation which generates the training process is generic. Since the differential equation can be identified with the vector field defined by its right side, it is sufficient to show genericity of a certain type of vector field.

Lemma 11.2 *There exists an open and dense set of vector fields which is contained in $\mathcal{G}$ such that the cascade ϕ_T, that is a discretization of the generated flow, is $\mathcal{T}$ robust. Let $\psi_{h,p}$ denotes the diffeomorphism generated by a Runge–Kutta method of order p and stepsize h which is applied to the equation generating the flow ϕ. Then, for each $p \in \{1, 2, ...\}$ and a sufficiently large m, the cascade $\Psi := \psi_{\frac{T}{n},p}^m$ is $\mathcal{T}$ robust as well.*

Proof Let $MS_{\mathcal{G}}$ denotes the family of all Morse-Smale vector fields contained in $\mathcal{G}$. Let us recall that $\mathcal{G} \subset \Gamma$ is the set of all vector fields of the form $-grad\ E$, where $E : \mathcal{M} \to \mathbb{R}, E \in C^2$. It is known that the family $MS_{\mathcal{G}}$ is open and dense in $\mathcal{G}$ - see for [148], p. 153.

Let us notice that if $-grad\ E$ belongs to $MS_{\mathcal{G}}$, then the critical points of the flow ϕ are also the fixed points of the cascade ϕ_T which has not any other fixed points. Furthermore, ϕ_T, like the flow ϕ, does not have other periodic points but the fixed points. Furthermore, both the stable and unstable manifolds of the flow ϕ and the cascade ϕ_T at their (common) fixed points are the same. Thus, ϕ_T is a Morse-Smale diffeomorphism and, by Lemma 5.7.8, is $\mathcal{T}$ robust, where $\mathcal{T} = \Theta_c \cup \Theta_s$. It can be easily verified that the vector field $-grad\ E \in MS_{\mathcal{G}}$ satisfies all the assumptions of Theorem 5.6.11. Thus, the cascades ϕ_T and Ψ are topologically conjugate if m

is sufficiently large. Furthermore, Theorem 5.7.9 implies the robustness of Ψ. This completes the proof of Lemma 11.2 and, as the consequence, the proof of Theorem 11.1.   □

Let us briefly comment on the presented idea.

A specific $n-$dimensional manifold $\mathcal{M}_S^n$, homeomorphic to the sphere $\mathcal{S}^n$, has been constructed. The subset of the manifold is geometrically identical with the subset of the Euclidean space $\mathbb{R}^n$ and the learning process of the perceptron is, in practice, analysed only on this subset. Therefore, the dynamical system $\left(\mathcal{A} = B^n(\mathbf{0}, 2r_1), \tilde{\phi} \right)$ that models a perceptron training process remains unchanged after transforming the problem onto the manifold. More formally, the theory applied to the analysis of the training process of perceptrons concerns the properties of gradient flows and cascades on compact manifolds without boundaries. The numerical scheme, however, which is implemented as the training algorithm, is performed in $\mathbb{R}^n$. Therefore, the conclusions should concern directly the set that is a domain of the implemented algorithm. This set, let us denote it as $\mathcal{A} \subset \mathbb{R}^n$, has to satisfy the following relations:

$$B^n(\mathbf{0}, r_1) \subset \mathcal{A} \subset Base \quad \text{and} \quad \alpha_m(\mathcal{A}) \subset Base$$

- see Steps 1 and 2 of the proof of Theorem 11.1

Let us analyse properties of the set $\alpha_m(B^n(\mathbf{0}, 2r_1))$. First of all, the following condition is satisfied: $\alpha_m(B^n(\mathbf{0}, r_1)) \subset \alpha_m(B^n(\mathbf{0}, 2r_1))$. Furthermore, since the conjugating homeomorphism α_m converges to identity for m converging to infinity - see Theorem 5.6.11, the following condition is also satisfied: $\alpha_m(B^n(\mathbf{0}, 2r_1)) \subset Base$, if only m is sufficiently large. Thus, topological conjugacy of the considered cascades exists on the set $\mathcal{A} = B^n(\mathbf{0}, 2r_1) \subset \mathbb{R}^n$ on which the training process of a perceptron is implemented.

The introduced construction allowed us to apply Theorem 5.6.11 as the theoretical basis of the performed analysis and, as the consequence, it has been shown that the dynamical system that models the training process is, under some natural assumptions, correctly reproduced by its Runge–Kutta method of each order if only a single step of the numerical method is sufficiently small. The dynamics of gradient flows is very regular. In particular, the system has no periodic orbits and the dynamics cannot be chaotic. It implies asymptotic stability of each training process, that is based on any Runge–Kutta method, and it ensures satisfying the stop condition of the training algotithm. These properties are preserved by the cascades generated by every Runge–Kutta method because of the existence of global topological conjugacy between the numerical scheme and the time-h-map flow discretization. It also implies $\mathcal{T}$ robustness of the training process.

In the presented theory the robustness was considered for $k \in \mathbb{Z}$. In the implemented algorithms, however, it is interested in practice only for $k \in \mathbb{N}$. The ball $B^n(\mathbf{0}, 2r_1)$ is a positively invariant set according to the cascade $\hat{\phi}_T$ and robustness is a topological conjugacy invariant - see Theorem 5.7.9. Therefore, according to the above conclusion concerning topological conjugacy, the robustness with respect to $\mathbb{N}$ takes place on the ball $B^n(\mathbf{0}, 2r_1)$.

Let us notice that the discrete dynamical system Ψ, that models the training process of perceptrons, is a multi-step operator. It should be stressed that the term *multi-step* should not lead to the confusion with a multi-step discretization method. Thus, the m-fold iteration of the operator Ψ, generated by the applied Runge–Kutta method is considered as a single unit of the theoretical analysis. In practice, it does not produce any limitation because during implementations we can consider the results of training process after each m-step stage.

It should be noticed that the presented results have one drawback. Namely, the implemented numerical schema are only piecewise continuous and therefore they are not contained in the considered classes of δ-methods. In order to fill this gap it should be proved that the aforementioned learning process ϕ_T or $\psi_{\frac{T}{m},p}$ is inverse shadowing with respect to the class generated by numerical methods

$$\left\{ \psi_{\frac{T}{m},p,b}, \ b \in \{1, 2, \ldots\} \right\},$$

where the index b is connected with the round-off with the set up accuracy, let us say 2^{-b}. Such methods are piecewise constant, so the presented theory is insufficient because of the presence of the points of discontinuity.

To sum up, some properties of the training process of the perceptrons can be analyzed by using dynamical systems theory. The training algorithms, that are described by numerical schema applied to formula (11.1), and the gradient dynamical systems generated by the training process are, generically, convergent to equilibrium states and they are robust. It implies, among others, accuracy, stability and well defined stop condition when they are implemented as computer algorithms. Nevertheless, the complementary studies of the inverse shadowing with respect to piecewise continuous methods are necessary to complete the obtained results.

11.1 Bibliographic Remarks

In this chapter the results described in [29, 46, 47] are presented.

The mentioned biological aspects concerning liberation of neurotransmitters from the presynaptic bouton to synaptic cleft - see [91] - Sect. 2.5 and [176] - pp. 39–40.

The construction of the cut-off function, used in the first step of the proof of Theorem 11.1, is described in details in [123], Lemma 2.21.

Chapter 12
Concluding Remarks and Comments

In this monograph three topics, mutually complementary, are studied:

- modelling of the neuron,
- modelling of the processes that take place in the neuron,
- mathematical analysis of dynamical properties of gradient training processes of perceptrons.

There are various reasons for presentation the topic in such perspective. First of all, there are structures and processes in the neuron that have not been modelled yet. Let us put an example. In the presynaptic bouton which has an irregular shape there are inner structures, for instance - mitochondrion. In all the models of the synaptic processes created up till now, in which geometry of the bouton was taken into consideration, the bouton was assumed to be a ball and the presence of the inner structures was neglected [37, 114]. Such assumptions are extremely simplifying. The models in which the bouton shape is based on microscopic imaging are more realistic but they are at the very preliminary stage of development [38]. In such models the inner structures are taken into consideration, as well. The author intended to signalize the problem and give a preliminary review of the structures and processes that can be modelled. Therefore, in Chaps. 2 and 7, especially in Sect. 7.3, biological foundations have been discussed in detail. Secondly, the review of the models of the neuron and its parts is given. On the one hand, there are several models of functional aspects of fragments of the neuron but, so far, they have not been put together. There are at least two reasons. The high computational complexity of numerical realization of such models is the first one. The doubts whether such a complex model of the neuron could contribute crucially new aspects in comparison with far simpler models of the whole neuron is the second reason. Nevertheless, it should be stressed that it has not been proved that such approach cannot contribute new aspects - this problem should be investigated carefully. On the other hand, although there are several functional models of the whole neuron, only few of them are used in artificial neural networks. To sum up, if the aforementioned levels of analysis are considered, i.e.

© Springer International Publishing AG, part of Springer Nature 2019
A. Bielecki, *Models of Neurons and Perceptrons: Selected Problems and Challenges*, Studies in Computational Intelligence 770,
https://doi.org/10.1007/978-3-319-90140-4_12

- the level of modelling of sub-neural structures and processes;
- the level of modelling of functional properties of the whole neuron;
- the level of designing and implementing ANNs,

then it can be easily noticed that only a tiny part of the potentiality of the lower level, i.e. the more detailed one, is utilized at the higher level. Thus, presentation of this problem and, as a consequence, recommending the inter-level studies together with pointing out their directions are some of the novelty aspects in this monograph. It is even more significant because, usually, the topics of the researchers' interests are included inside one of the single levels specified above. The synthetic approach to the modelling of neural processes is the second aspect of this monograph specificity. In this monograph the cybernetic modelling, the mathematical modelling and the modelling by using electronic circuits intertwine. It is clear, especially, in Sect. 7.3. This is also a specificity of the approach presented in this monograph because, usually, these three ways of modelling are exploited separately. The problem of the synthetic approach to the scientific studies is more general. Since the Enlightenment analytic approach to scientific problems has dominated and the synthetic approach is, in general, in the state of atrophy. The synthetic mathematical-electronic approach to modelling of sub-neural processes, presented in this monograph, is a test whether such approach can be efficient. The results show that the answer is affirmative. The presentation of the consistent model of the perceptron structure and training process dynamics is the subsequent topic of this monograph. The analysis of the mathematical properties of the dynamics of the training process are crucial part of these studies. The approach which consists in application of dynamical systems theory to analysis of the training process of ANNs in not the branch of studies that is exploited intensively. Nevertheless, apart from the results presented in this monograph, some other interesting studies have been conducted - the papers [73–75] can be put as examples.

Part V
Appendix

Chapter 13
Approximation Properties of Perceptrons

As it was mentioned in Sect. 8.1, an untrained perceptron can be treated as a family of functions $\mathbb{R}^n \to \mathbb{R}^m$ indexed by a vector set of all its weights. A given training set, in turn, can be regarded as a set of the points to which a mapping should be approximated in the best way. The investigations of approximation abilities of neural networks are focused on the existence of an arbitrarily close approximation. They are also focused on the problem how accuracy depends on a complexity of a perceptron. In this chapter a few basic theorems that concern the approximation properties of perceptrons are discussed. The presented theorems are the classical results. In this monograph they are presented without the proofs which can be found in literature.

Kolmogorov theorem on the representation of continuous function of many variables by superposition of continuous functions of one variable is the main basis of the studies concerning perceptrons in the context of approximation. Let us recall it.

Theorem 13.1 (Kolmogorov Approximation Theorem) *For each natural $n \geq 2$ there exists a family of functions $\{\psi_{pq}\}$, $p \in \{1, \ldots, n\}, q \in \{1, \ldots, 2n + 1\}$, continuous on the interval $[0, 1]$, such that on the $n-$dimensional cube I^n any continuous function $f : I^n \to \mathbb{R}$ can be represented in the following form*

$$f(x_1, \ldots x_n) = \sum_{q=1}^{2n+1} \chi_q \left(\sum_{p=1}^{n} \psi_{pq}(x_p), \right),$$

where $\{\chi_q\}$, $q \in \{1, \ldots, 2n + 1\}$ is a family of continuous real functions.

On the basis of Kolmogorov Approximation Theorem Hecht–Nielsen proposed a perceptron with one hidden layer as a system for approximation a continuous function $f : I^n \longrightarrow \mathbb{R}$. The proposed perceptron consisted of n input units that put the input signal $\mathbf{x} = [x_1, \ldots, x_n]$ onto the hidden layer that consisted of $2n + 1$ neurons. The output signals of the neurons in the hidden layer were of the following form

© Springer International Publishing AG, part of Springer Nature 2019
A. Bielecki, *Models of Neurons and Perceptrons: Selected Problems and Challenges*, Studies in Computational Intelligence 770,
https://doi.org/10.1007/978-3-319-90140-4_13

$$u_q = \sum_{p=1}^{n} \lambda_{pq} \psi_q(x_p),$$

where $\{\lambda_{pg}\}$, $q \in \{1, \ldots, 2n+1\}$, $p \in \{1, \ldots, n\}$ was a family of real constants. The output layer consisted of a single neuron which generated the output signal

$$y = \sum_{q+1}^{2n+1} \phi(u_q),$$

where ϕ is a continuous nonlinear function.

The above proposal has only existential character because the way of the construction of functions ψ_q and ϕ remained unknown. Cybenko presented a more constructive result. Let us consider a perceptron with one hidden layer which consists of N neurons. Each one has an increasing continuous activation function $f : \mathbb{R} \to \mathbb{R}$ that satisfies the following limit conditions

$$\lim_{s \to -\infty} f(s) = 0, \quad \lim_{s \to \infty} f(s) = 1 \tag{13.1}$$

The neurons had a threshold γ_k, $k \in \{1, \ldots, N\}$. The output layer consisted of a single linear neuron. Input signals were $d-$dimensional i.e. $\mathbf{x} \in \mathbb{R}^d$. The perceptron output was given by the formula

$$y_N(\mathbf{x}) = \sum_{k=1}^{N} v_k \cdot f(\mathbf{w_k} \circ \mathbf{x} + \gamma_k), \tag{13.2}$$

where $\mathbf{w_k}$ is a vector of weights of the $k-$th neuron of the hidden layer, whereas $\mathbf{v} = [v_1, \ldots, v_N]$ is a vector of weights of the output neuron. Such a perceptron can approximate, with an arbitrary accuracy, any continuous function on a compact set.

Theorem 13.2 (Cybenko Approximation Theorem) *Let a function f, that satisfies the conditions specified above, be given. Then, for any continuous function $g : I^d \to \mathbb{R}$ and each $\varepsilon > 0$, there exists a natural number N and parameters $\mathbf{v}$, $\mathbf{w_k}$, γ_k such that for any $\mathbf{x} \in I^d$*

$$|g(\mathbf{x}) - y_N(\mathbf{x})| < \varepsilon.$$

In Cybenko theorem the activation functions are given explicitly but the theorem gives neither a way of parameter choice nor approximation accuracy. The last one was given by Barron for the perceptrons described by formula (13.2) with increasing activation function which satisfies the conditions (13.1). In order to present this estimation of approximation let us specify two following assumptions that concern a function $g : \mathbb{R}^d \to \mathbb{R}$:

C1 The function g has an integral representation of the form

$$g(\mathbf{x}) = \int_{\mathbb{R}^d} e^{i \cdot \omega^T \cdot \mathbf{x}} \cdot \tilde{g}(\omega) d\omega,$$

where $\tilde{g}$ is a function of a complex variable.

C2 The function $\sqrt{\omega^T \omega} \cdot \tilde{g}(\omega)$ is integrable in $\mathbb{R}^d$ which implies that

$$C_f := \int_{\mathbb{R}^d} \sqrt{\omega^T \cdot \omega} \cdot |\tilde{g}(\omega)| d\omega < \infty.$$

Let $\Im_d$ denote the family of functions $g : \mathbb{R}^d \longrightarrow \mathbb{R}$ which satisfies the conditions C1 and C2.

Theorem 13.3 *Let us assume that a function $f : \mathbb{R} \to \mathbb{R}$ is the activation function which satisfies the conditions specified above that concern the activation function f. Let us choose a ball $\mathbb{R}^d \supset B_r := \{\mathbf{x} \in \mathbb{R}^d : ||\mathbf{x}|| < r, r > 0\}$. Let us also assume that $g \in \Im_d$. Then, for each natural $N \geq 1$, there exist such parameters $\mathbf{v} = [v_1, \ldots, v_N]$, $\mathbf{w_k}, \gamma_k$, that the perceptron (13.2) realizes a function $y_N : \mathbb{R}^d \to \mathbb{R}$ which satisfies the inequality*

$$\int_{B_r} (g(\mathbf{x}) - y_N(\mathbf{x}))^2 < \frac{2 \cdot r \cdot C_f}{N}.$$

The theorem can be easily generalized. First of all, the approximation domain can differ from a ball. Secondly, the measure which is not a Lebesgue one can be used.

13.1 Bibliographic Remarks

Kolmogorov approximation theorem is proved in [116].

Hecht–Nielsen approximator is described in [87].

Cybenko theorem is presented in [59].

The proof of Barron theorem [11], i.e. Theorem 11.2, is based on Strong Law of Large Numbers.

The other theorems and their proofs related to the problem of approximation properties of the perceptron can be found in [88, 96–98, 117].

Chapter 14
Proofs

Two proofs that are not commonly known, are presented in this chapter.

14.1 Estimation of Constants in Fečkan Theorem

In this section the proof of Theorem 5.6.8, conducted by Jabłoński, is presented - see [39, 101]. In this section, both linear operators and corresponding matrices are denoted by capital letters. The matrices, however, are not denoted in bold because the operator aspect of the considered objects is crucial.

The following lemma is the starting point.

Lemma 14.1.1 *Let* $\varepsilon > 0$ *be given. Let, furthermore,* $\phi_1, \phi_2 \in C_B(\mathbb{R}^n)$ *be Lipshitzean mappings with the Lipschitz constants less than* ε. *Let us also assume that* $A \in \mathcal{L}(\mathbb{R}^n)$ *is an endomorphism without eigenvalues on the imaginary axis. Then, for* $0 < h < \|A\|^{-1}$, *the mappings* $A_h + \phi_1$ *and* $A_h + \phi_2$ *are topologically conjugate provided that*

$$\frac{\varepsilon \|A_h^{-1}\|}{1 - M_h} < 1, \tag{14.1}$$

where $M_h = \max \{\|(A_h^u)^{-1}\|, \|A_h^s\|\}$.

In order to prove Lemma 14.1.1 the following lemma is needed - see [148], p. 60, Lemma 4.3:

Lemma 14.1.2 *Let* $G, K \in \mathcal{L}(\mathbb{R}^n)$. *There exists* $\mu \in [0, 1)$ *such that* $\|K\|, \|G^{-1}\|$ $\leq \mu$ *and*

(i) *the mapping* $I + K$ *is automorphism and* $\|(I + K)^{-1}\| \leq \frac{1}{1-\mu}$,
(ii) *the mapping* $I + G$ *is automorphism and* $\|(I + G)^{-1}\| \leq \frac{\mu}{1-\mu}$.

© Springer International Publishing AG, part of Springer Nature 2019
A. Bielecki, *Models of Neurons and Perceptrons: Selected Problems
and Challenges*, Studies in Computational Intelligence 770,
https://doi.org/10.1007/978-3-319-90140-4_14

Proof of Lemma 14.1.1 First a mapping $s : \mathbb{R}^n \to \mathbb{R}^n$ of the form $s = I + u$, where $u \in \mathcal{C}_B(\mathbb{R}^n)$, will be constructed. The mapping s has to satisfy the following equality

$$(A_h + \phi_1) \circ s = s \circ (A_h + \phi_2). \tag{14.2}$$

The Eq. (14.2) can be written in the form

$$A_h \circ u - u \circ (A_h + \phi_2) = \phi_2 - \phi_1 \circ (I + u). \tag{14.3}$$

It will be shown that there exists a unique mapping $s_0 \in \mathcal{C}_B(\mathbb{R}^n)$ which satisfies Eq. (14.3).

Let us define a linear operator

$$\Lambda : \mathcal{C}_B(\mathbb{R}^n) \longrightarrow \mathcal{C}_B(\mathbb{R}^n)$$

in the following way

$$\Lambda(u) = A_h \circ u - u \circ (A_h + \phi_1).$$

Since the matrix A has not imaginary eigenvalues and eigenvalues of A_h are different from 1, the operator Λ is invertible and, by Lemma 14.1.2

$$\|\Lambda^{-1}\| \leq \|A_h^{-1}\| \cdot (1 - M_h)^{-1}.$$

Let us consider the mapping

$$\zeta : \mathcal{C}_B(\mathbb{R}^n) \longrightarrow \mathcal{C}_B(\mathbb{R}^n)$$

defined as follows

$$\zeta(u) = \Lambda^{-1} \circ (\phi_2 - \phi_1 \circ (I + u)).$$

Let us estimate distance between $\zeta(u_1)$ and $\zeta(u_2)$ for any $u_1, u_2 \in \mathcal{C}_B(\mathbb{R}^n)$.

$$\|\zeta(u_1) - \zeta(u_2)\| = \|\Lambda^{-1} \circ [\phi_2 \circ (I + u_2) - \phi_1 \circ (I + u_1)]\| \leq$$

$$\leq \varepsilon (1 - M_h)^{-1} \cdot \|A_h^{-1}\| \cdot \|u_1 - u_2\|.$$

The above inequality implies that ζ is a contraction provided that

$$\varepsilon < (1 - M_h) \cdot \|A_h^{-1}\|^{-1}.$$

By Banach Fixed Point Theorem, there exists a unique mapping

$$u_0 \in \mathcal{C}_B(\mathbb{R}^n)$$

which satisfies (14.3). This completes the proof of Lemma 14.1.1.

The following lemma was proved in [175, Theorem 1.8.1].

Lemma 14.1.3 *Let* $F(h, x) = e^{hA}x$, $x \in \mathbb{R}^n$ *and let* $\Phi(h, x)$ *be the time-h-map of the flow generated by (5.8). Then the mappings* $\Phi(h, \cdot)$, $F(h, \cdot)$ *are topologically conjugate provided that*

$$hb \cdot \|e^{Ah}\| \cdot (\|A\| + b) < (1 - M) \cdot \|A_h^{-1}\|^{-1}.$$

Proof of Lemma 14.1.3 For a fixed $x \in \mathbb{R}^n$ the equation

$$\frac{d\Phi(t, x)}{dt} = A\Phi(t, x) + g(\Phi(t, x)),$$

is, on the interval $[0, h]$, equivalent to the integral equation

$$\Phi(t, x) = e^{tA}x + f(t, x),$$

where

$$f(t, x) = \int_0^t e^{(t-\tau)A} g(\Phi(\tau, x)) d\tau.$$

Since $\|Dg(x)\| \leq b$ for each $x \in \mathbb{R}^n$,

$$\|D\Phi(x)\| = \|A + Dg(x)\| \leq \|A\| + b.$$

The derivative $Df : \mathbb{R}^n \to \mathcal{L}(\mathbb{R}^n)$ is bounded:

$$\|Df(x)\| \leq \int_0^t \|e^{(t-\tau)A}\| \cdot \|Dg(\Phi(t, x))\| \cdot \|D\Phi(x)\| d\tau \leq$$

$$\leq \int_0^t b \cdot \|e^{tA}\| \cdot (\|A\| + b) d\tau = tb \cdot \|e^{tA}\| \cdot (\|A\| + b).$$

For $t = h$ the mapping $f(h, \cdot)$ is Lipschitzean with the constant

$$\varepsilon = hb \cdot \|e^{Ah}\| \cdot (\|A\| + b).$$

By Lemma 14.1.1, the mappings $\Phi(h, \cdot)$, $F(h, \cdot)$ are topologically conjugate if

$$\varepsilon < (1 - M_h) \cdot \|A_h^{-1}\|^{-1}.$$

This implies

$$hb \cdot \|e^{Ah}\| \cdot (\|A\| + b) < (1 - M_h) \cdot \|A_h^{-1}\|^{-1}$$

which means that the inequality (5.12) is proved.

Lemma 14.1.4 *Let the mappings* $F(h, x) = e^{hA}x$ *and* $G(h, x) = A_h x$. *The mappings* $F(h, \cdot)$ *and* $G(h, \cdot)$ *are topologically conjugate.*

Proof The number of elements of the set $\{v \in \sigma(e^{hA}) \mid v > 1\}$ is equal both to the number of elements of the set $\{v \in \sigma(A_h) \mid v > 1\}$ and to the number of the eigenvalues less than 1. That means the cascades F and G have the same index and, as the consequence, are topologically conjugate.

Lemma 14.1.5 *Let* $b > 0$ *be given. Let the mapping* $g \in C_B^1(\mathbb{R}^n)$ *be such that*

$$\|Dg(x)\| < b \ \text{ for each } \ x \in \mathbb{R}^n.$$

Then, the mappings $G(h, \cdot)$, $H(h, \cdot)$ *are topologically conjugate provided that*

$$hb < (1 - M) \cdot \|A_h^{-1}\|^{-1}.$$

Proof Since $\|Dg(x)\| \leq b$ for each $x \in \mathbb{R}^n$, the mapping hg is Lipschitzean with a constant $\varepsilon = hb$. By Lemma 14.1.1 the mappings $G(h, \cdot)$ and $H(h, \cdot)$ are topologically conjugate provided that

$$hb \, (1 - M_h)^{-1} < \|A_h^{-1}\|^{-1},$$

which completes the proof of Lemma 14.1.5.

Inequality (5.11) is implied by Lemma 14.1.1. This completes the proof of Theorem 5.6.8.

14.2   Estimation of the Euler Method Error on a Manifold

Let $\mathcal{M}$ be a k−dimensional Riemanian C^j manifold, $j \geq 2$, embedded in $\mathbb{R}^{2k+1}$ - compare Whithey Theorem 4.5. For the points $x, y \in \mathcal{M}$, transformed by the same chart ϑ, the following inequalities hold (see [160]), p. 453, formula (2.2))

$$m_1 \cdot d_{\mathbb{R}^k}(\vartheta(x), \vartheta(y)) \ \leq \ \varrho_{Riem}(x, y) \ \leq \ m_2 \cdot d_{\mathbb{R}^k}(\vartheta(x), \vartheta(y)), \qquad (14.4)$$

where m_1, m_2 are constant for a given map ϑ.

Let the time step h be constant. By the inequalities (14.4), for sufficiently small h, the Euler method on compact subset of a manifold is, similarly as in Euclidean space, a method of the first order.

The error of a single step of the Euler method is a continuous function of x. Let us denote by

$$r(x, h) := \varrho_R(\psi_h(x), \phi_h(x)) \qquad (14.5)$$

the error of a single step of the Euler method.

The set $\mathcal{M} \supset \mathcal{A} := \{y \in \mathcal{M}, \; \varrho_R(\phi(x_0, t), y) \le e_n, \; t \in T\}$ is compact.
Let us define

$$r_n(h) := \max_{m \le n} \left\{ r\left(\phi_h^m(x_0), h\right)\right\}.$$

Then

$$e_n(x_0, \tilde{x}_0, h) := \varrho_R(\; \psi_h^n(x_0), \; \phi_h^n(\tilde{x}_0)\;) \;\; = \varrho_R\left(\; \psi_h(\psi_h^{n-1}(x_0)), \; \phi_h(\phi_h^{n-1}(\tilde{x}_0))\;\right) \;\; \le$$

$$\varrho_R\left(\; \psi_h(\psi_h^{n-1}(x_0)), \; \psi_h(\phi_h^{n-1}(\tilde{x}_0))\;\right) \;\; + \;\; \varrho_R\left(\; \psi_h(\phi_h^{n-1}(\tilde{x}_0)), \; \phi_h(\phi_h^{n-1}(\tilde{x}_0))\;\right) \;\; \le$$

$$\varrho_R\left(\; \psi_h(\psi_h^{n-1}(x_0)), \; \psi_h(\phi_h^{n-1}(\tilde{x}_0))\;\right) + r_n(h).$$

By formula (5.3) the last inequality can be written as follows, by using local exponents

$$e_n(x_0, \tilde{x}_0, h) \le$$

$$\le \varrho_R\left(\; \exp_{\psi_h^{n-1}(x_0)}\left(h \cdot f(\psi_h^{n-1}(x_0))\right), \; \exp_{\phi_h^{n-1}(\tilde{x}_0)}\left(h \cdot f(\phi_h^{n-1}(\tilde{x}_0))\right)\;\right) + r_n(h).$$

Let $\mathcal{W}$ be an envelope of the bundle TM zero section defined as $TM(0) := \{(p, 0), \; p \in \mathcal{M}\} \subset TM$. The mapping $\exp : \mathcal{W} \to \mathcal{M}$ is defined as $\exp(p, \cdot) = \exp_p$. The exponent mapping can be constructed on every C^j manifold if $j \ge 2$ and it is a C^{j-1} map. The restriction of the exponent to the zero section is the identity mapping. By using the exponent mapping instead of the local exponents the error is estimated in the following way

$$e_n(x_0, \tilde{x}_0, h) \le$$

$$\varrho_R\left(\; \exp(\psi_h^{n-1}(x_0), \; h \cdot f(\psi_h^{n-1}(x_0))), \; \exp(\phi_h^{n-1}(\tilde{x}_0), \; h \cdot f(\phi_h^{n-1}(\tilde{x}_0)))\;\right) + r_n(h).$$

The maximal difference between the values of the function in two different points is upper bounded by the product of the maximum value of the function differential on the interval determined by these points and the interval length. Since $j \ge 2$, the differential of the exponent is, at least, a C^1 mapping. Thus, it is a Lipschitz mapping on a compact space. As the set $\mathcal{A}$ is compact, the envelope $\{(p, \mathbf{v}), p \in \mathcal{A}, \mathbf{v} \in T_pM, \|\mathbf{v}\| \le \varepsilon\}$ of $TM(0)$ is also compact. Let $\mathcal{U}_h := \{ (p, \mathbf{v}), \; p \in \mathcal{M}, \; \|\mathbf{v}\| \le h \cdot L_1 \}$ be a small closed envelope $\mathcal{U}_h$ of the set $\mathcal{A} \subset TM(0)$. Let us notice that the envelope "width" is proportional to the time step h. Let, furthermore, for every point $p \in \mathcal{A}$, the pair $(p, h \cdot f(p))$ be contained in $\mathcal{U}_h$. Since the envelope $\mathcal{U}_h$ is compact, the exponent differential is a Lipschitz mapping and its norm on the zero section is equal to 1. The maximum of the differential norm can be estimated by $(1 + h \cdot c)$, where c is a constance depending on L_1. Thus

$$e_n(x_0, \tilde{x}_0, h) \le$$

$$\leq (1 + h \cdot c) \cdot \varrho_{TM} \left((\psi_h^{n-1}(x_0),\; h \cdot f(\psi_h^{n-1}(x_0))),\; (\phi_h^{n-1}(\tilde{x}_0),\; h \cdot f(\phi_h^{n-1}(\tilde{x}_0))) \right) + r_n(h),$$

where ϱ_{TM} is a metric on the tangent bundle consistent with its topology. This metric can be defined as follows

$$\varrho_{TM}((p_1, \mathbf{v}_1),\; (p_2, \mathbf{v}_2)) := d_{\mathbb{R}^k}(p_1, p_2) + \| \mathbf{v}_1 - \mathbf{v}_2 \|_{\mathbb{R}^k},$$

where $\| \cdot \|_{\mathbb{R}^k}$ is a norm on $\mathbb{R}^k$. Thus, the error can be estimated in the following way

$$e_n(x_0, \tilde{x}_0, h) \leq$$

$$(1 + h \cdot c) \cdot \left[d_{\mathbb{R}^k} \left(\psi_h^{n-1}(x_0),\, \phi_h^{n-1}(\tilde{x}_0) \right) + \| h \cdot f(\psi_h^{n-1}(x_0)) - h \cdot f(\phi_h^{n-1}(\tilde{x}_0)) \|_{\mathbb{R}^k} \right] + r_n(h)$$

$$= (1 + h \cdot c) \cdot \left[d_{\mathbb{R}^k} \left(\psi_h^{n-1}(x_0),\, \phi_h^{n-1}(\tilde{x}_0) \right) + h \cdot \| f(\psi_h^{n-1}(x_0)) - f(\phi_h^{n-1}(\tilde{x}_0)) \|_{\mathbb{R}^k} \right] + r_n(h).$$

The metrics ϱ_R and $d_{\mathbb{R}^k}$ are equivalent on compact sets. Therefore, the map f is also a Lipschitz function on the set $\mathcal{A}$ considered as a subset of $\mathbb{R}^k$

$$e_n(x_0, \tilde{x}_0, h) \leq$$

$$\leq (1 + h \cdot c) \cdot \left[d_{\mathbb{R}^k} \left(\psi_h^{n-1}(x_0),\, \phi_h^{n-1}(\tilde{x}_0) \right) + h \cdot L_2 \cdot d_{\mathbb{R}^k} \left(\psi_h^{n-1}(x_0),\, \phi_h^{n-1}(\tilde{x}_0) \right) \right] + r_n(h)$$

$$= (1 + h \cdot c) \cdot (1 + h \cdot L_2) \cdot d_{\mathbb{R}^k} \left(\psi_h n - 1(x_0),\, \phi_h^{n-1}(\tilde{x}_0) \right) + r_n(h).$$

The length of the line segment which connects two points in $\mathbb{R}^n$ is less than the length of any curve connecting the same points, thus $d_{\mathbb{R}^k}(p, \tilde{p}) \leq \varrho_{Riem}(p, \tilde{p})$. Therefore,

$$e_n(x_0, \tilde{x}_0, h) \leq (1 + h \cdot c) \cdot (1 + h \cdot L_2) \cdot \varrho_R \left(\psi_h^{n-1}(x_0),\, \phi_h^{n-1}(\tilde{x}_0) \right) + r_n(h) =$$

$$= (1 + h \cdot c) \cdot (1 + h \cdot L_2) \cdot e_{n-1} + r_n(h).$$

Iteratively,

$$e_n(x_0, \tilde{x}_0, h) \leq (1 + h \cdot c) \cdot (1 + h \cdot L_2) \cdot [(1 + h \cdot c) \cdot (1 + h \cdot L_2) \cdot e_{n-2} + r_n(h)]$$
$$+ r_n(h) =$$

$$= [(1 + h \cdot c) \cdot (1 + h \cdot L_2)]^2 \cdot e_{n-2} + [(1 + h \cdot c) \cdot (1 + h \cdot L_2)]$$
$$\cdot r_n(h) + r_n(h) = \ldots$$

$$[(1+h \cdot c) \cdot (1+h \cdot L_2)]^n \cdot e_0 + r_n(h) \cdot \{ 1 + [(1+h \cdot c) \cdot (1+h \cdot L_2)]$$
$$+ [(1+h \cdot c) \cdot (1+h \cdot L_2)]^2 + \ldots$$

$$\ldots + [(1+h \cdot c) \cdot (1+h \cdot L_2)]^{n-1} \} =$$

$$[(1+h \cdot c) \cdot (1+h \cdot L_2)]^n \cdot e_0(x_0, \tilde{x}_0)$$
$$+ r_n(h) \cdot \frac{(1+h \cdot c)^n \cdot (1+h \cdot L_2)^n - 1}{(1+h \cdot c) \cdot (1+h \cdot L_2) - 1} =$$

$$= [(1+h \cdot c) \cdot (1+h \cdot L_2)]^n \cdot e_0(x_0, \tilde{x}_0)$$
$$+ r_n(h) \cdot \frac{(1+h \cdot c)^n \cdot (1+h \cdot L_2)^n - 1}{1 + h \cdot (c + L_2) + h^2 \cdot c \cdot L_2 - 1} <$$

$$< [(1+h \cdot c) \cdot (1+h \cdot L_2)]^n \cdot e_0(x_0, \tilde{x}_0)$$
$$+ r_n(h) \cdot \frac{(1+h \cdot c)^n \cdot (1+h \cdot L_2)^n - 1}{h \cdot (c + L_2)}.$$

The Euler method is a numerical method of the first order, thus $r_n(h) \leq b \cdot h^2$. Since for $s > 0$ the inequality $(1 + s \cdot h) < e^{s \cdot h}$ is satisfied and $a = n \cdot h$, it is obtained

$$e_n(x_0, \tilde{x}_0, h) \leq e^{n \cdot h \cdot c} \cdot e^{n \cdot h \cdot L_2} \cdot e_0(x_0, \tilde{x}_0) + b \cdot h \cdot \frac{e^{n \cdot h \cdot c} \cdot e^{n \cdot h \cdot L_2} - 1}{c + L_2} =$$

$$= e^{a \cdot (c+L_2)} \cdot e_0(x_0, \tilde{x}_0) + b \cdot h \cdot \frac{e^{a \cdot (c+L_2)} - 1}{c + L_2}.$$

Thus

$$e_n(x_0, \tilde{x}_0, h) \leq e^{a \cdot L} \cdot e_0(x_0, \tilde{x}_0) + \frac{b}{L} \cdot \left(e^{a \cdot L} - 1 \right) \cdot h.$$

References

1. Abdul, R. A., Alabi, A. R. A., & Tsien, R. W. (2012). Synaptic vesicle pools and dynamics. *Cold Spring Harbor Perspectives in Biology, 4,* 2276–2278.
2. Agogino, A., Tseng, M-L., & Jain, P. (1993). *Integrating neural networks with influence diagrams for power plant monitoring and diagnostics: Neural Network Computing for the Electric Power Industry: Proceedings of the 1992 INNS (International Neural Network Society) Summer Workshop,* 213–216.
3. Aihara, K., Takabe, T., & Toyoda, M. (1990). Chaotic neural networks. *Physical Letters, Series A, 144,* 333–340.
4. Amit, D. (1989). *Modeling brain functions.* Cambridge: Cambridge University Press.
5. Aristizabal, F., & Glavinovic, M. I. (2004). Simulation and parameter estimation of dynamics of synaptic depression. *Biological Cybernetics, 90,* 3–18.
6. Ashby, W. R. (1948). Design for a brain. *Electronic Engineering, 20,* 379–383.
7. Ashby, W. R. (1957). *An introduction to cybernetics.* London: Chapman & Hall LTD.
8. Ashby, W. R. (1962). Principles of self-organizing system. In H. Von Foerster & G.W. Zopf Jr. (Eds.) *Principles of self-organization: transactions of the university of Illinois symposium* (255–278). Pergamon Press: London. (also in: Emergence: Complexity and Organizations, vol.6, 2004, 102–126, section: Classical Works.)
9. Atkins, C. M., & Sweatt, J. D. (1999). Reactive oxygen species mediate activity-dependent neuron-glia signaling in output fibers of the hippocampus. *Journal of Neuroscience, 19,* 7241–7248.
10. Azoff, E. M. (1994). *Neural network time series forecasting of financial markets.* Chichester: Wiley.
11. Barron, A. R. (1993). Universal approximation bounds for superposition of sigmoidal functions. *IEEE Transactions on Information Theory, 39,* 930–944.
12. Barszcz, T., Bielecka, M., Bielecki, A., & Wójcik, M. (2011). Wind turbines states classification by a fuzzy-ART neural network with a stereographic projection as a signal normalization. *Lecture Notes in Computer Science, 6594,* 225–234.
13. Barszcz, T., Bielecki, A., Bielecka, M., Wójcik, M. & Włuka, M. (2016). Vertical axis wind turbine states classification by an ART-2 neural network with a stereographic projection as a signal normalization. In Chaari et al. (Eds.) *Advances in condition monitoring of machinery in non-stationary operations.* Applied condition monitoring (Vol. 4, pp. 265–275).
14. Barszcz, T., Bielecki, A., Wójcik, M. & Bielecka, M. (2014). ART-2 artificial neural networks applications for classification of vibration signals and operational states of wind turbines for intelligent monitoring. In *Advances in condition monitoring of machinery in non-stationary operations.* Lecture notes in mechanical engineering (pp. 679–688).

© Springer International Publishing AG, part of Springer Nature 2019

A. Bielecki, *Models of Neurons and Perceptrons: Selected Problems and Challenges*, Studies in Computational Intelligence 770,
https://doi.org/10.1007/978-3-319-90140-4

15. Barszcz, T., Bielecki, A., & Wójcik, M. (2010). ART-type artificial neural networks applications for classification of operational states in wind turbines. *Lecture Notes in Artificial Intelligence, 6114*, 11–18.

16. Baxt, W. G. (1990). Use of an artificial neural network for data analysis in clinical decision-making: The diagnosis of coronary occlusion. *Neural Computation, 2*, 480–489.

17. Baxt, W. G. (1991). Use of an artificial neural network for the diagnosis of myocardial infarction. *Annals of Internal Medicine, 165*, 843–848.

18. Bąk, M. & Bielecki, A. (2006). *Methodology of neural systems development*. In Cader A., Rutkowski L., Tadeusiewicz R. & Żurada J. (Eds.) Artificial intelligence and soft computing. Challenging problems of science - computer science, Bolc L. - the series editor (pp. 1–7). Warszawa: Academic Publishing House EXIT.

19. Bąk, M., & Bielecki, A., (2007). Neural systems for short-term forecasting of electric power load. *Lecture Notes in Computer Science, 4432*, 133–142.

20. Beltratti, A., Margarita, S., & Terna, P. (1996). *Neural networks for economic and financial modelling*. London: International Thomson Compute Press.

21. Bennett, M. V. L. (2000). Seeing is relieving: Electrical synapses between visualized neurons. *Nature Neuroscience, 3*, 7–9.

22. Bennett, M. V. L., & Zukin, R. S. (2004). Electrical coupling and neuronal synchronization in the mammalian brain. *Neuron, 41*, 495–511.

23. Bianchini, R. (1983). Local controllability, rest states and cyclic points. *SIAM Journal on Control and Optimization, 21*, 714–720.

24. Bianchini, R. (1983). Instant controllability of linear autonomous systems. *Journal of Optimization Theory and Applications, 39*, 237–250.

25. Bielecka, M., Bielecki, A., Korkosz, M., Skomorowski, M., Wojciechowski, W., & Zieliński, B. (2011). Modified Jakubowski shape transducer for detecting osteophytes and erosions in finger joints. *Lecture Notes in Computer Science, 6594*, 147–155.

26. Bielecka, M., Bielecki, A., Korkosz, M., Skomorowski, M., Wojciechowski, W., & Zieliński, B. (2010). Application of shape description methodology to hand radiographs interpretation. *Lecture Notes in Computer Science, 6374*, 11–18.

27. Bielecka, M., Skomorowski, M., & Zieliński, B. (2009). A fuzzy shape descriptor and inference by fuzzy relaxation with application to description of bones contours at hand radiographs. *Lecture Notes in Computer Science, 5495*, 469–478.

28. Bielecki, A. (2000). Topological conjugacy of cascades generated by gradient flows on the two-dimensional sphere. *Annales Polonici Mathematici, 73*, 37–57.

29. Bielecki, A. (2001). Dynamical properties of learning process of weakly nonlinear and nonlinear neurons. *Nonlinear Analysis: Real World Applications, 2*, 249–258.

30. Bielecki, A. (2002). Estimation of the Euler method error on a Riemannian manifold. *Communications in Numerical Methods in Engineering, 18*, 757–763.

31. Bielecki, A. (2002). Topological conjugacy of discrete time-map and Euler discrete dynamical systems generated by a gradient flow on a two-dimensional compact manifold. *Nonlinear Analysis, 51*, 1293–1317.

32. Bielecki, A. (2003). Mathematical model of architecture and learning process of artificial neural networks. *Task Quarterly, 7*, 93–114.

33. Bielecki, A. (2015). A general entity of life - a cybernetic approach. *Biological Cybernetics, 109*, 401–419.

34. Bielecki, A. (2017). *Epistemological meaning of mathematical models and computer simulations in subcellular biology,* Semina Scientiarum, 16, (in Polish).

35. Bielecki, A., Barszcz, T., Wójcik, M., & Bielecka, M. (2014). Hybrid system of ART and RBF neural networks for classification of vibration signals and operational states of wind turbines. *Lecture Notes in Artificial Intelligence, 8467*, 3–11.

36. Bielecki, A., Bielecka, M., & Chmielowiec, A. (2008). Input signals normalization in Kohonen neural networks. *Lecture Notes in Artificial Intelligence, 5097*, 3–10.

37. Bielecki, A., Gierdziewicz, M. & Kalita, P. (2017). *Three-dimensional simulation of presynaptic release and depression,* under revision.

38. Bielecki, A., Gierdziewicz, M., Kalita, P., & Szostek, K. (2016). Construction of a 3D geometric model of a presynaptic bouton for use in modeling of neurotransmitter flow. *Lecture Notes in Computer Science, 9972*, 377–386.
39. Bielecki, A., & Jabłoński, D. (2002). Estimation of numerical dynamics constants of a weakly nonlinear neuron. *Lecture Notes in Computer Science, 2328*, 862–869.
40. Bielecki, A., Jabłoński, D., & Kędzierski, M. (2004). Properties and applications of weakly nonlinear neurons. *Journal of Computational and Applied Mathematics, 164–165*, 93–106.
41. Bielecki, A., & Kalita, P. (2008). Model of neurotransmitter fast transport in axon terminal of presynaptic neuron. *Journal of Mathematical Biology, 56*, 559–576.
42. Bielecki, A., & Kalita, P. (2012). Stability, controllability and observability properties of the averaged model of fast synaptic transport. *Journal of Mathematical Analysis and Applications, 393*, 329–240.
43. Bielecki, A., Kalita, P., Lewandowski, M., & Skomorowski, M. (2008). Compartment model of neuropeptide synaptic transport with impulse control. *Biological Cybernetics, 99*, 443–458.
44. Bielecki, A., Kalita, P., Lewandowski, M., & Siwek, B. (2010). Numerical simulation for neurotransmitter transport model in axon terminal of presynaptic neuron. *Biological Cybernetics, 102*, 489–502.
45. Bielecki, A., Korkosz, M., & Zieliúski, B. (2008). Hand radiographs preprocessing, image representation in the finger regions and joint space width measurements for image interpretation. *Pattern Recognition, 41*, 3786–3798.
46. Bielecki, A., & Ombach, J. (2004). Shadowing property in analysis of neural networks dynamics. *Journal of Computational and Applied Mathematics, 164–165*, 107–115.
47. Bielecki, A., & Ombach, J. (2011). Dynamical properties of a perceptron learning process: Structural stability under numerics and shadowing. *Journal of Nonlinear Science, 21*, 579–593.
48. Bielecki, A., Podolak, I. T., Wosiek, J., & Majkut, E. (1996). Phonematic translation of Polish texts by the neural network. *Acta Physica Polonica, Series B, 27*, 2253–2264.
49. Bielecki, A., Skomorowski, M. & Woźniak, R. (2006). *Electronic neural networks modelling*, In Cader A., Rutkowski L., Tadeusiewicz R. & Żurada J. (Eds.) Artificial intelligence and soft computing. Challenging problems of science - computer science, Bolc L. - series editor (pp. 8–14). Warszawa: Academic Publishing House EXIT.
50. Bobrowski, A., & Morawska, K. (2012). From a PDE model to an ODE model of dynamics of synaptic depression. *Discrete and Continuous Dynamical Systems B, 17*, 2313–2327.
51. Boyd, I. A., & Martin, A. R. (1956). The end-plate potential in mammalian muscle. *Jourenal of Physiology, 132*, 74–91.
52. Brammer, R. F. (1972). Controllability in linear autonomous systems with positive controllers. *SIAM Journal on Control, 10*, 339–353.
53. Broomhead, S., & Lowe, D. (1988). Multivariable functional interpolation and adaptive network. *Complex Systems, 2*, 321–323.
54. Chen, C. T. (1970). *Introduction to linear systems theory*. New York: Holt, Rinehart and Winston Inc.
55. Chen, C. T., & Desoer, C. A. (1967). Controllability and observability of composite systems. *IEEE Transactions on Automatic Control, 12*, 402–409.
56. Chen, L., & Aihara, K. (1995). Chaotic simulated annealing by a neural network model with transient chaos. *Neural Networks, 8*, 915–930.
57. Corless, R., & Pilyugin, S Yu. (1995). Approximate and real trajectories for generic dynamical systems. *Journal of Mathematical Analysis and Applications, 189*, 409–423.
58. McCulloch, W. S., & Pitts, W. H. (1943). A logical calculus of the ideas immanent in nervous activity. *Bulletin of Mathematical Biophysics, 5*, 115–133.
59. Cybenko, G. (1989). Approximation by superposition of a sigmoidal function. *Mathematics of Control, Signals and Systems, 2*(1989), 303–314.
60. Dauer, J. P. (1971). Perturbations of linear control systems. *SIAM Journal on Control, 9*, 393–400.
61. Demidowicz, B. (1972). *Mathematical Theory of Stability*. Warszawa: WNT (in Polish).

62. del Castillo, J., & Katz, B. (1954). Quantal components of the end-plate potential. *Journal of Physiology, 124,* 560–573.

63. Destexhe, A., Mainen, Z. F., & Sejnowski, T. (1994). Synthesis of models for excitable membranes, synaptic transmission and neuromodulation using a common kinetic formalism. *Journal of Computational Neuroscience, 1,* 195–230.

64. Dreyfus, S. (1973). The computational solution of optimal control problems with time lag. *IEEE Transactions on Automatic Control, 18*(4), 383–385.

65. Du, Y., Wang, F. & Cheng, T. C. (1993). *A case study of neural network application: Power equipment failure diagnosis,* In *Neural Network Computing for the Electric Power Industry: Proceedings of the 1992 INNS (International Neural Network Society) Summer Workshop* (pp. 207–211).

66. Evans, L. C. (1998). *Partial differential equations.* Rhode Island: American Mathematical Society.

67. Fečkan, M. (1992). The relation between a flow and its discretization. *Mathematica Slovaca, 42,* 123–127.

68. Fečkan, M. (2011). *Bifurcation and Chaos in discontinuous dynamical systems.* Berlin: Springer.

69. Fields, R. D. (2004). The other half of the brain. *Scientific American, 290*(4), 54–61.

70. Fields, R. D., & Burnstock, G. (2006). Purinergic signalling in neuron - glia interactions. *Neture Reviews, 7,* 422–436.

71. Fields, R. D., & Stevens, B. (2000). ATP: An extracellular signaling molecule between neurons and glia. *Trends in Neuroscience, 23,* 556–562.

72. Fields, R. D., & Stevens-Graham, B. (2002). New insights into neuron-glia communication. *Science, 298,* 556–562.

73. Fiori, S. (2011). Extended Hamiltonian learning on Riemannian manifolds: Theoretical aspects. *IEEE Transactions on Neural Networks, 22,* 687–700.

74. Fiori, S. (2012). Extended hamiltonian learning on Riemannian manifolds: Numerical aspects. *IEEE Transactions on Neural Networks and Learning Systems, 23,* 7–21.

75. Fiori, S., Kaneko, T., & Tanaka, T. (2012). Learning on the compact Stiefel manifold by a Cayley-transform-based pseudo-retraction map. *In Proceedings of the International Joint Conference on Neural Networks, 2012,* 3829–3832.

76. FitzHugh, R. (1961). Impulses and physiological states in theoretical models of nerve membrane. *Biophysical Journal, 1,* 445–466.

77. FitzHugh, R. (1969). Mathematical models of excitation and propagation in nerve. In H. P. Schwann (Ed.), *Biological engineering* (pp. 1–85). New York: McGraw-Hill.

78. Flasiński, M. (2016). *Introduction to artificial intelligence.* Berlin: Springer.

79. Galarreta, M., & Hestrin, S. (1999). A network of fast-spiking cells in the neocortex connected by electrical synapses. *Nature, 402,* 72–75.

80. Garay, B. (1993). Discretization and some qualitative properties of ordinary differential equations about equilibria. *Acta Mathematica Universitatis Comenianae, 62,* 245–275.

81. Garay, B. (1994). Discretization and Morse-Smale dynamical systems on planar discs. *Acta Mathematica Universitatis Comenianae, 63*(1994), 25–38.

82. Garay, B. (1996). On structural stability of ordinary differential equations with respect to discretization methods. *Journal of Numerical Mathematics, 4,* 449–479.

83. Garay, B. (1996). On C^j -closeness between the solution flow and its numerical approximation. *Journal of Difference Equations and Applications, 2,* 67–86.

84. Garay, B. (1998). The discretized flow on domains of attraction: A structural stability result. *IMA Journal of Numerical Analysis, 18,* 77–90.

85. Greengard, P. (2001). The neurobiology of slow synaptic transmission. *Science, 294,* 1024–1030.

86. Harmon, L. D. (1961). Studies with artificial neurons, I: Properties and functions of an artificial neuron. *Kybernetik, 1,* 89–101.

87. Hecht-Nielsen, R. (1987). *Kolmogorov's mapping neural network existence theorem.* In *Proceedings of the International Conferencion Neural Networks, Part III* New York: IEEE Press.

88. Hecht-Nielsen, R. (1990). *Neurocomputing*. Reading: Addison-Wesley Publ.
89. Hebb, D. O. (1949). *The organization of behavior*. New York: Wiley.
90. Hertz, J., Krogh, A., & Palmer, R. G. (1991). *Introduction to the theory of neural computation*. Massachusetts: Addison-Welsey Publishing Company.
91. Hess, G. (2009). *Synaptic transmission and synaptic plasticity*. In Tadeusiewicz R. (Ed.) Theoretical Neurocybernetics. Warsaw: Warsaw University Press (in Polish).
92. Hestrin, S., & Galaretta, M. (2005). Electrical synapses define networks of neocortical GABAergic neurons. *Trends in Neurosciences, 28,* 304–309.
93. Heymann, M., & Stern, P. (1975). Controllability of linear systems with positive controls: Geometric considerations. *Journal of Mathematical Analysis and Applications, 52,* 36–41.
94. Hodgkin, A. L., & Huxley, A. F. (1952). A quantitative description of membrane current and its application to conduction and excitation in nerve. *Journal of Physiology, 117,* 500–544.
95. Hopfield, J. (1984). Neurons with graded response have collective computational properties like those of two-states neurons. *Proceedings of National Academy of Science (USA), 81,* 3088–3092.
96. Hornik, K. (1991). Approximation capabilities of multilayer feedforward networks. *Neural Networks, 4,* 251–258.
97. Hornik, K. (1993). Some new results on neural network approximation. *Neural Networks, 6,* 1069–1072.
98. Hornik, K., Stinchcombe, M., & White, H. (1989). Mulitilayer feedforward networks are universal approximators. *Neural Networks, 2,* 359–366.
99. Irwin, M. C. (1087). *Smooth dynamical systems*. London: Academic Press Ltd.
100. Izhikevich, E. M. (2007). *Dynamical systems in neuroscience. The geometry and excitability and bursting*. In Sejnowski T. J., Poggio T. A. (Eds.) Computational Neuroscience. MIT Press
101. Jabłoński, D. (2002). The conjugacy between cascades generated a perturbated linear system and the Euler method of a flow. *Applicationes Mathematicae, 29,* 43–49.
102. Jabłoński, D. & Bielecki, A. (2000). *Numerical conditions for the Fečckan Theorem and applications to the artificial neural networks*. In *Proceedings of the Sixth National Conference on Application of Mathematics in Biology and Medicine, Zawoja* p. 5760.
103. Jack, J. J. B., Noble, D., & Tsien, J. W. (1975). *Electric current flow in excitable cells*. Oxford: Oxford University Press.
104. Kaczorek, T. (1974). *Theory of automatic control systems*. Warszawa: WNT. (in Polish).
105. Keener, J. P. (1983). Analog circuitry vor the van der Pol and FitzHugh-Nagumo equation. *IEEE Transactions on Systems, man and Cybernetics.*
106. Keener, J., & Sneyd, J. (1998). *Mathematical physiology, Interdisciplinary applied mathematics*. New York: Springer.
107. Kępiński, A. (1972). *Melancholy, Wydawnictwo Literackie, Kraków, 2001 (in Polish)* (1st ed.). Warszawa: PZWL.
108. Kim, C., In-keun, Y., & Song, Y. H. (2002). Kohonen neural network and wavelet transform based approach to short-term load forecasting. *Electric Power Systems Research, 63,* 169–176.
109. Klamka, J. (1991). *Controllability of Dynamical Systems (co-edition) Warszawa: PWN Polish Scientific Publishers*. Dordrecht/Boston/London: Kluwer Academic Publishers.
110. Kleinle, J., K. Vogt, K., Lüscher, H. R. L, Müller, L., Senn, W., Wyler, K. & Streit J. (1996). *Transmitter concentration profiles in the synaptic cleft: An analytical model of release and diffusion. Biophysical Journal, 71,* 2413–2426.
111. Kloeden, P. E., & Lorenz, J. (1986). Stable attracting sets in dynamical systems and their one-step discretizations. *SIAM Journal on Numerical Analysis, 23,* 986–996.
112. Kloeden, P. E., & Ombach, J. (1997). Hyperbolic homeomorphisms are bishadowing. *Annales Polonici Mathematici, 65,* 171–177.
113. Kloeden, P. E., Ombach, J., & Pokrowskii, A. (1999). Continuous and inverse shadowing. *Journal of Functional Differential Equations, 6,* 135–151.
114. Knodel, M. M., Geiger, R., Ge, L., Bucher, D., Grillo, A., Wittum, G., Schuster, C. & Queisser, G. (2014). Synaptic bouton properties are tuned to best fit the prevailing firing pattern. *Frontiers in Computational Neuroscience, 8,* article 101.

115. Koch, C., & Segev, I. (Eds.). (1989). *Methods in neuronal Modeling.* Cambridge, MA: MIT Press.
116. Kolmogorov, A. N. (1957). On the representation of continuous function of many variables by superposition of continuous functions of one variable. *Doklady Akademii Nauk SSSR, 114,* 953–956.
117. Kurkova, V. (1992). Kolmogorov's theorem and multilayer neural networks. *Neural Networks, 5,* 501–506.
118. Lang, S. (1999). *Fundamentals of differential geometry.* Berlin: Springer.
119. Lawrence, J. (1994). Designing back propagation neural networks: A financial predictor example. *Neurowe$T Journal, 2*(1), 8–13.
120. LeDoux, J. (1994). Emotion, memory and the brain. *Scientific American, 270*(6), 50–57.
121. LeDoux, J. (1996). *The emotional brain.* New York: New York Times Company.
122. Lee, E., & Markus, L. (1968). *Foundations of optimal control theory.* New York: Academic Press.
123. Lee, J. M. (2003). *Introduction to smooth manifolds.* New York: Springer.
124. Levine, S., Finn, C., Darrell, T., & Abbeel, P. (2016). End-to-end training of deep visuomotor policies. *Journal of Machine Learning Research, 17,* 1–40.
125. Levitan, E. S. (2008). Signaling for vesicle mobilization and synaptic plasticity. *Molecular Neurobiology, 37,* 3943.
126. Li, M. C. (1997). Structural stability of Morse-Smale gradient-like flows under discretization. *SIAM Journal of Mathematical Analysis, 28,* 381–388.
127. Li, M. C. (1997). Structural stability of flows under numerics. *Journal of Differential Equations, 141,* 1–12.
128. Li, M. C. (1999). Structural stability of the Euler method. *SIAM Journal on Mathematical Analysis, 30,* 747–755.
129. Li, M. C. (1999). Structural stability on basins for numerical methods. *Proceedings of the American Mathematical Society, 127,* 289–295.
130. Li, M. C. (2003). Stability of parametrized Morse-Smale gradient-like flows. *Discrete and Continuous Dynamical Systems, 9,* 1073–1077.
131. Madden M. G., O'Connor N. (2005). *Neural network approach to predicting stock exchange movements using external factors.* In *Proceedings of the 25th International Conference on Innovative Techniques and Applications of Artificial Intelligence AI-2005.* Cambridge.
132. Mass, W. (1997). Networks of spiking neurons: The third generation of neural network models. *Neural Networks, 10,* 1659–1671.
133. McKean, H. P. (1970). Nagumo's equation. *Advances in Mathematics, 4,* 209–223.
134. Mazur, M. (1966). *Cybernetic theory of autonomous systems.* Warszawa (in Polish): PWN.
135. Mazur, M. (1976). *Cybernetics and character.* Warszawa (in Polish): PIW.
136. Medved, M. (1983). On genericity of complete controllability in the space of linear parametrized control systems. *Czechoslovak Mathematical Journal, 33,* 167–175.
137. Mikia, Y., Muramatsua, C., Hayashib, T., Zhoua, X., Haraa, T., Katsumatac, A., et al. (2017). Classification of teeth in cone-beam CT using deep convolutional neural network. *Computers in Biology and Medicine, 80,* 2429.
138. Minsky, M., & Papert, S. (1969). *Perceptrons.* Cambridge: MIT Press.
139. Müller, B., & Reinhardt, J. (1990). *Neural networks.* New York: Springer.
140. Moody, D., & Darken, J. (1989). Fast learning in networks of locally-tuned processing units. *Neural Computation, 1,* 281–294.
141. Nagumo, J., Arimoto, S., & Yoshizawa, S. (1964). An active pulse transmission line simulating nerve axon. *Proceedings of the IRE, 50,* 2061–2070.
142. Nicoll, R. A., & Alger, B. E. (2004). The brains own marijuana. *Scientific American, 291*(6), 68–75.
143. Ogiela, M. & Tadeusiewicz, R. (2008). *Modern computational intelligence methods for the interpretation of medical images.* Studies in Computational Intelligence, vol. 84. New York: Springer.

144. Ogiela, M., Tadeusiewicz, R., & Ogiela, L. (2006). Image languages in intelligent radiological palm diagnostics. *Pattern Recognition, 39*, 21572165.
145. Ombach, J. (1993). The simplest shadowing. *Annales Polonici Mathematici, 58*, 243–158.
146. Ombach, J. & Mazur, M. (2002). Shadowing and likes as C^0 generic properties. In Kryszewski, W. & Nowakowski, A. (Eds.) *Proceedings of the 3rd Polish Symposium on Nonlinear Analysis.* Lecture notes in nonlinear analysis, vol.3, Juliusz Schauder Center for Nonlinear Studies, Nicholas Copernicus University, pp. 159–189.
147. Osowski, S., & Siwek, K. (2002). Regularization of neural networks for improved load forecasting in the power system. *IEE Proceedings of Generation, Transmission and Distribution, 149*, 340–344.
148. Palis, J., & de Melo, W. (1982). *Geometric theory of dynamical systems.* New York: Springer.
149. Parzen, E. (1962). On estimation of a probability density function and mode. *Annals of Mathematical Statistics, 33*, 1065–1076.
150. Pasemann, F. (1993). Dynamics of a single model neuron. *International Journal of Bifurcation and Chaos, 3*, 271–278.
151. Perez Velazquez, J. L. (2005). Brain, behaviour and mathematics: are we using the right approaches? *Physica D, 212*, 161–182.
152. Pilyugin, S. Y. (1999). *Shadowing in dynamical systems.* Lecture Notes in Mathematics, 1706. New York: Springer.
153. Podolak, I. T. (1998). Functional graph model of a neural network, IEEE Trans. on Systems. *Men and Cybernetics - Part B: Cybernetics, 28*, 876–881.
154. Podolak, I. T. (1999). Feedforward neural network's sensitivity to input data representation. *Computer Physics Communications, 117*, 181–188.
155. Podolak, I. T., & Bielecki, A. (2003). A neural system of phonematic transformation. *Task Quarterly, 7*, 115–130.
156. Podolak, I. T., Lee, S., Bielecki, A., Majkut, E. (2000). *A hybrid neural system for phonematic transformation.* In *Proceedings of the 15-th International Conference on Pattern Recognition IAPR, Barcelona* (pp. 961–964).
157. Rall, W. (1977). *Core conductor theory and cable properties of neurons.* In Brookhart, J. M., Mountcastle, V. B. (Eds.) Handbook of physiology. The nervous system. Bethesda, MD: American Physiological Society.
158. Refenes, A. P., & Zaidi, A. (1995). Managing exchange-rate prediction strategies with neural networks. In A. P. Refenes (Ed.), *Neural networks in the capital markets.* New York: Wiley.
159. Rizzoli, S. O., & Jahn, R. (2007). Kiss-and-run, collapse and 'eadily retrievable' vesicles. *Traffic, 8*, 1137–1144.
160. Robbin, J. (1971). A structural stability theorem. *Annals of Mathematics, 94*, 447–493.
161. Rosslenbroich, B. (2009). The theory of increasing autonomy in evolution: a proposal for understanding macroevolutionary innovations. *Biology and Philosophy, 24*, 623–644.
162. Rosslenbroich, B. (2014). *On the origin of autonomy. A new look at the major transitions in evolution.* In Wolfe, C. T., Huneman, P., Reydon, T. A. C. (Eds.) History, philosophy and the life sciences, vol. 5. Berlin: Springer
163. Ruiz-Mirazo, K., & Moreno, A. (2012). Autonomy in evolution: from minimal to complex life. *Synthese, 185*, 21–52.
164. Rumelhart, D. E., Hinton, G. E., & Williams, R. J. (1986). Learning representations by back-propagating errors. *Nature, 323*(6088), 533–536.
165. Schmitendorf, W., & Barmish, B. (1980). Null controllability of linear system with constrained controls. *SIAM Journal on Control and Optimization, 18*, 327–345.
166. Scott, A. (1995). *Stairway to the mind.* New York: Springer.
167. Sejnowski, T. J., & Rosenberg, C. R. (1987). Parallel networks that learn to pronounce English text. *Complex Systems, 1*, 145–168.
168. Shakiryanova, D., Tully, A., Hewes, R. S., Deitcher, D. L., & Levitan, E. S. (2005). Activity-dependent liberation of synaptic neuropeptide vesicles. *Nature Neuroscience, 8*, 173–178.
169. Shakiryanova, D., Tully, A., & Levitan, E. S. (2006). Activity-dependent synaptic capture of transiting peptidergic vesicles. *Nature Neuroscience, 9*, 896–900.

170. Sheperd, G. M. (1983). *Neurobiology*. New York - Oxford: Oxford University Press.
171. Sheperd, G. M. (1996). The dendritic spine: a multifunctional integrative unit. *Journal of Neurophysiology, 75*, 2197–2210.
172. Specht, D. F. (1990). Probabilistic neural networks. *Neural Networks, 3*, 109–118.
173. Swadlow, H. A., Kocsis, J. D., & Waxman, S. G. (1980). Modulation of impulse conduction along the axonal tree. *Annual Review of Biophysics and Bioengineering, 9*, 143–179.
174. Sykova, E. (1997). The extracellular space in the CNS: Its regulation, volume and geometry in normal and pathological neuronal function. *Neuroscientist, 3*, 28–41.
175. Szlenk, W. (1984). *An Introduction to the theory of smooth dynamical systems PWN*. Warsaw, John Wiley and Sons Inc, New York: Polish Scientific Publishers.
176. Tadeusiewicz, R. (1994). *Problems of Biocybernetics*. Warszawa (in Polish): PWN.
177. Tadeusiewicz, R. & Ogiela, M. (2004). *Medical image understanding technology*. Studies in fuzziness and soft computing, vol. 156. Berlin: Springer
178. Takefuji, Y., & Lee, K. C. (1991). An artificial hysteresis binary neuron; a model suppresing the oscillatory behoviours of neural dynamics. *Biological Cybernetics, 64*(1991), 353–356.
179. Tuckwell, H. C. (1988). *Introduction to theoretical neurobiology*. Cambridge, New York: Cambridge University Press.
180. Waxman, S. G. (1972). Regional differentiation of the axon: A review with special reference to the concept of the multiplex neuron. *Brain Research, 47*, 269–288.
181. Werbos, P. J. (1975). *Beyond regression: new tools for prediction and analysis in the behavioral sciences*. Harvard University.
182. Wiener, N. (1947). *Cybernetics or control and communication in the animal and the machine*. Cambridge: MIT Press.
183. Wiggins, S. (1990). *Introduction to applied nonlinear dynamical systems and chaos*. New York: Springer.
184. Wightman, R. M., & Haynes, C. L. (2004). Synaptic vesicles really do kiss and run. *Nature Neuroscience, 7*, 321–322.
185. Wilson, R. I., & Nicoll, R. A. (2002). Endocannabinoid signaling in the brain. *Science, 296*, 678–682.
186. Yoon, Y. O., Brobst, R. W., Bergstresser, P. R., Peterson, L. L. (1989). A desktop neural network for dermatology diagnosis. *Journal of Neural Network Computing, Summer*, 43-52.
187. Zirilli, J. S. (1996). *Financial prediction using neural networks*. London: International Thomson Computer Press.
188. Żurada, J., Barski, M., & Jędruch, W. (1996). *Artificial Neural Networks*. PWN, Warszawa: (in Polish).

Printed in the United States
By Bookmasters